FIZZICKS and CAUSEMOLOGY

...Considerations Given Without The Complicated Mathematics...

JOHN BEAUPRE

PAGE PUBLISHING
Conneaut Lake, PA

First originally published by Page Publishing 2023

ISBN 978-1-6624-6416-4 (pbk)
ISBN 978-1-6624-6418-8 (hc)
ISBN 978-1-6624-6417-1 (digital)

Printed in the United States of America

Contents

Preface

Why "Fizzicks and Cosmology" at All?

How is it best to try to understand the most basic questions?

How are we here? What and where <u>is</u> here?

Why are we here? What <u>is</u> our purpose?

Philosophy (with lots of answers) and religion (with lots of unanswered questions) may be more entertaining pursuits. But, once we, *Homo sapiens*, change our collective minds and change our collective behaviors, so then our philosophies must (and our religions tend to) change accordingly. "Scientology" is a rather recent religion, or <u>is</u> it actually a philosophy...for example? Who can say for sure?

Our sciences may mature, but do not and cannot change...and still remain as sciences. Specific scientific understandings, like those within physics and cosmology in particular, are and must remain solid and unchanging. Force has always been and always will be the product of mass times acceleration. Gravity is constant too, and Newton's equation will never change. Nor will the speed of light, nor the ratio of any circumference to its diameter, nor the mass (nor size nor charge) of a proton. Because...

This is our space...but <u>where</u> is it?

This is our time...but <u>when</u> is it?

This book cannot and will not answer these questions exactly. But, its observations will get us closer to whatever understandings and answers are currently available. And...

Mathematics...for other than a very specialized type of person...is (are) confusing. So the 'math' itself of the science within is minimized herein. Logic, however, is recognizable and is understandable...and is employed throughout this book. So...

Do enjoy the 'read'...as I have enjoyed the 'write.'

John Beaupré

Foreword

In June 1984, I wound up in New Orleans, Louisiana. My flight to France and ultimately to Biarritz for a bicycle tour to Barcelona had been canceled…for the day. So I had a day (my first one) to explore New Orleans. Accordingly…

My plan was first to visit the city's highly regarded aquarium, there on Decatur Street. It was a bit of a walk from my hotel on Bourbon Street…and I was wearing my favorite, old, well-used Western cowboy boots. The walk was 'healthy,' so to stay tuned for the ride, but it was also slower in my old boots. I had purchased them at "Cowboyz" in Santa Fe for $125 one year earlier. And *only* I knew that.

En route, I came upon two young boys just 'hang'n out' by the waterfront there on Decatur. One of the boys, obviously the 'older and wiser' of the two, stopped me and said, "Hey, man, I know where you got dem boots." (Pardon me if you've heard a story like this before, but understand that I had not back then in 1984.)

And I took the bait. I knew for a fact that boy *did not* know where (in Santa Fe) I had purchased (at Cowboyz) my old boots. So suspecting a hustle of some sort from the kids, I replied, "How much? How much will you bet that you know where I got my boots?"

The older boy's reply was, "Five bucks." And the hustle was on!

I pulled out a Lincoln (their term when they saw my $5 bill) and, a bit like a smart aleck, waved it in their faces. I said, "Okay. I'll take that bet. But where's *your* five bucks?"

The lead kid replied, "My friend here has it, and he'll hold yours too."

I replied, "And you and he will run away with my money...right?"

"No, no," they replied. "We'll both put our fives here on the curb...no tricks, and we'll tell you where you got dem boots yer wearin'."

I figured I was in good enough shape and was quick enough to grab my five if they'd 'split.' And even if not, I was willing to see this through if they ran away. After all, I was confident in my bet and liked this kid's style. The whole exchange was worth five bucks anyway.

The next minute or two seemed more like an hour... and was well worth it. The older boy moved into the best 'moonwalk' I'd ever seen, better even than Michael Jackson's...and did so while the smaller kid beat out a rhythm on his thighs. They both performed a kind of very early 'rap' tune for me:

> Ya got dem boots on yer feet, and
> Ya got yer feet on Decatur Street, and
> Dat's where ya got dem boots.

I'd been had—clearly! I'd lost the bet, fair and square!

So I picked up *my* five before they could, and they essentially and in tandem both responded, "C'mon, man, we told you right...yes?"

I replied, "Yep, you did...and good for you." Then I replaced my $5 bill with a $10 bill and asked them, "Please do that one more time. It was worth it. You moonwalk better than Jackson, and you're right—I do 'got' my feet in my boots here on your street!"

They repeated their routine. I paid. They smiled. And I asked, "On a good day, how many times do you guys do this?"

They replied, "'Bout twenty on a really good day... kinda like t'day."

I figured (to myself): *Say a hundred dollars plus cash per day, say five to seven hundred, tax-free per week is not bad for a couple of enterprising twelve- or fourteen-year-old boys...here in New Orleans. Good for them!*

Again, so what? Why now, after almost thirty-seven years, do I (or can I) even bother to remember this?

Answer: It was a perfectly logical and symmetrical (both sides won) and enjoyable and memorable event at a specific point in time, and it occurred at a very perfect place...visited once only. And the experience, for $10, was definitely worth it. The memory of it is stored in my 'gray matter,' along with the moment I asked Sharon to marry me, *and* she said "yes" (at the LA airport, Trans World Airlines arrival gate in 1968) and the moment that I won my first ski race (actually, first *run* of a slalom) at White Pass, Washington. I recall that finish in particular and still have my blue ribbon...but forgot the date! Anyway, they all remain special points and places in time. Note also that Sharon and I have been married now for almost fifty-five years!

Question: what does all this have to do with science? Answer: Everything. And this would be...

'Close enough' to what we can at best recall...perhaps it's a bit like quantum mechanics? And this explains...what? QM has been, to date, always 'close enough,' but this needs to be discussed a bit later...possibly in *Fizzicks 201*. Back to 1984, and if confused, the sense of it all is <u>where</u> I was and <u>when</u> I was then…and that <u>that</u> confirms the point of my existence…both then and now.

Our ride across southern France and northern Spain, through the Pyrenees and across 'Basque-land' was great. The cycling and the roads were near perfect, and I repeated my $5 then $10 'dem boots' story to my friends several times. Then upon my way back to the United States…

For no particular reason, other than an enduring interest in science fiction at the time, I decided upon a movie when back in New Orleans. I found I had almost another day… actually, a ten-plus-hour layover. What to do? It was a short cab ride, and the theater ticket was $2.50 less than my total bets had been back on Decatur. A movie, any movie, would be a much more interesting choice than sitting in an airport for ten hours. The film I chose was *Buckaroo Banzai, Across the Eighth Dimension*. I saw it for the first time then, and yes, I've seen it several times since. It too, for a science fiction connoisseur, is definitely worth it. Peter Weller's (the lead actor's) best line as Buckaroo (the film's protagonist) was, as I recall it *almost* perfectly today:

Wherever ya go, there you are.

It was a take upon the logic of Confucius, and so, out of some curiosity, I checked further and the actual film clip confirmed it:

No matter where you go…there you are.

So what? The messages and logic were and are identical.

June of 1984 again… Ten-plus days earlier, I'd begun with my feet (in dem boots) on Decatur Street. Then I'd had my feet (primarily in my cycling shoes) clipped to my bike's pedals, both in France and in Spain. And then having flown a second time over the Atlantic Ocean, I was once again back in Louisiana in a theater watching Mr. Bonzai, and then would be headed back to the New Orleans airport, ultimately to be flown back to Albuquerque, New Mexico, where I'd drive back to Santa Fe. Most assuredly then…

Wherever I had gone, there I had been.

It occurred to me, as had been confirmed to me collectively by those enterprising boys and Confucius and then also by Buckaroo:

Ya gotta be <u>some</u>where…always.

If this seems too obvious upon which to spend much time thinking (or writing?), it <u>is</u> worth knowing this: One's location in one's space at any moment both defines and itself exists as one's <u>place</u> in one's time. After all, why *are* we here? What is our purpose? For us, any and all of us, to exist, what is the actual 'reason'?

First, there is (has to be) <u>space</u>—which is absolute necessity number 1. Space's absence would be the ultimate contradiction of <u>any</u> existence. Then also, there has to be (must always have been) <u>time</u>—which is absolute necessity number 2. Put them together—that is, recognize and/or experience them simultaneously—and there is <u>place</u>. And a place for everyone and for everything is mandatory. We do not live in some sort of colossal hologram.

Wherever ya go, there you are.

and

Ya gotta be somewhere.

<u>Actually…</u>

I'd started in Santa Fe, stopped over in New Orleans, lost $10 to the kids, flown to Paris, then to Biarritz, unpacked my bike, pedaled to Barcelona, repacked my bike, driven back to Paris, flown back to New Orleans, caught a cab, gone to a movie, caught a late plane to Albuquerque, and driven home…all in roughly twelve days of my time. I had, via just that trip and those experiences, utilized my time (about 1,036,800 seconds of it) and had experienced a lot of my very own 'places.' I was able to have done this because…

Because of what's already happened over time. So in the general order of occurrence and from day one:

- First there is (must have been) <u>space</u>. Then
- add <u>time</u>, and there was…
- <u>place</u>, which was first occupied by…
- <u>energy</u>, initially, as was our universe's creation via…
- the <u>Big Bang</u>, which over time cooled to create…
- <u>matter</u> (recall: E <u>does</u> equal mc²), which was first represented by…
- the original <u>particles</u>, which somehow…
- assumed <u>identities</u> (as in individual masses, with or without electrical charges but with spins and valences) and…then they…
- became organized as <u>atoms</u>, which in a very short time…
- formed <u>elements</u>, primarily two gases, being…
- <u>hydrogen</u>, first and foremost, and then…
- <u>helium</u>, secondly. And both of them, while just 'floating around' in an infant universe…

- experienced <u>gravity</u>, which caused them to…
- coalesce into <u>stars</u> that initiated fusion and…
- into <u>planets</u> that did not, but included one that accumulated…
- <u>water,</u> and an atmosphere that was close enough to its star so to experience sufficient…
- <u>heat</u> that when combined with…
- inorganic <u>substances</u> and <u>chemicals</u> (and the liquid water), they somehow reorganized themselves as…
- <u>organic</u> substances and <u>cells</u> that…
- incorporated into (or created?) <u>life</u>, which…
- was able to <u>reproduce</u> as…
- living <u>organisms</u> that have resulted in…
- <u>us</u>!

Got it? That's essentially how it all came down…to allow me to have taken my trip. Without any one of these specific events above having happened <u>first </u>and in the exact order as above, I would not have lost (spent?) $10 with those kids, nor cycled to Barcelona, nor would I have seen *Buckaroo Bonzai*. It's simple as that.

I'm very sure that I <u>did</u> take/make that trip. But how are we to know for sure what happened before, long before me, so that I could have done so? And the honest answers are the following:

"We don't know" but "We're workin' on it."

There's a lot of matter and energy left to explain and to experience out and about, and probably plenty well beyond what we can know and understand. So we basically 'guess.' We theorize and try to predict <u>what is here</u> and <u>what is there</u> and how it all <u>works</u>. And that is…

the physics (aka *Fizzicks*)

And also, we want to know (1) <u>how</u> and (2) <u>when</u> it all came to be and would like to know (3) <u>where</u>. And that is…

the cosmology (aka Causemology)

And that's why I've taken the time to write this book while, and at the same time(s), I've also tried to avoid the mathematics…so to retain—much like my trip and Buckaroo's common sense were—to make it simple…simple to understand.

The science(s), more specifically the currently existing mathematics, is (are) complicated and confusing. Here's hoping enough of it (the complicated math) has been avoided. Therefore…

Perhaps if I can sell enough books, I can take another trip. And if this, my math, is really simple, then…

I can start all over again!

Good reading to you all. Let me know if you enjoyed your time doing it.

Chapter 1

Simplicity—Mediocity—Complexity

Mediocity? <u>Not</u> medio<u>crity</u>. Medio<u>city</u> is logically halfway between simplicity and complexity. It would be a new word—my own.

We (those I can remember and I) began our 'formal educations' in kindergarten. Montessori and 'day schools' did not exist when we were age five.

In our kindergarten, we were not 'graded' per se. It was a "show up or not" sort of preschool. Not even "pass or fail." It was pure simplicity: Just show up or not. But I do remember those early rides to school. Mom's car smelled like damp cloth…or something close to it due to the very wet climate there in the Pacific Northwest.

First grade at age six was distinctly more structured. We had assigned seats and books, not just play areas and toys. There were report cards…and real 'friends.' Those real friends got report cards too. We all naturally (?) wanted <u>our</u> cards to be better than <u>their</u> cards. Even if grades were either "pass" or "not pass." "Fail," even back then, was considered demeaning…or whatever. Most everyone I recall "passed" anyway. 'Failing' first grade was not an option.

And our early education remained relatively simple in our little grade school in Medina, Washington. Plus, there was Margot McDonald; she was really pretty, but no one could explain the spelling (that should have been) Margo (without the *T*). Of note here: Margot M. wound up marrying Mitch Milias, a fraternity brother of mine at Stanford, years later. Some paths run near parallel over time. Ours did. We stay in touch.

Also in grade 1 was Johnny Meisnest. He was a bit different. He collected bugs back then…like every bug he could find. I don't recall Johnny M. past grade 3 or 4. As said: Some paths do and some do not run parallel. Ours did not.

But early on, grade school remained simple if only just to rate or, better, to have rated one's attendance…with a bit of "enjoys participation" and "relates with others" notations added. These provided some measure of our early educations. Simplicity prevailed nonetheless.

The first actual grades on take-home report cards came shortly thereafter. These grades, neither objective nor alphabetical, were instead rather subjective and limited to four 'judgments'; either 1 = Excellent or 2 = Good or 3 = Fair or 4 = Poor. There was no 5 = "Bad," like there had been no "fail" earlier. Grades 3, 4, and 5 actually <u>had</u> grades. One was judged either "excellent" or "good" or "fair" or "poor." "Excellence" was achievable, and "poorness" was to be avoided at all costs!

Grades 6, 7, and 8 were more difficult to define. Junior high school <u>was</u> finally defined as actual grades 7 and 8. So grade 6 was our last chance at "excellence." Grade 7 and up was the 'real deal.' Margot M. was there; Johnny M. was not. We then first received <u>letter</u> grades…so A, B, C, or D, and possibly F. Note: Grades had neither definitions nor translations, except for F, which clearly stood for "fail," also clearly a definition of *failure* and was again to be avoided at all costs.

So our pre-high school student rankings remained, for the first time, alphabetical, and simplicity had been replaced with mediocity. That is to say if one was paying attention, one was regarded as being either an A or B or C student. I don't recall that any D student rankings were assigned.

Those with Fs were (rather unceremoniously) held back to repeat grades 7 or 8. Being 'held back' was again to be avoided. Two 'goofball' kids—I recall their faces but not their names—simply disappeared...only to reappear later in different schools. One for sure reappeared and then enrolled at our Bellevue High School, but as a 'special needs' sort of student. A nice guy, but I assume he never graduated. However, he was a terrific linebacker and fullback on our football team...and 'started' all three years while I was there. Maybe he did so a few more after I graduated...it's hard to know for sure.

One might say that the letter grades were the transition between the simplistic (pass-fail) and more *mediocitic* (new word again) alphabetical grading. For sure, when letter grades became quantified, mediocity had been definitely achieved. Said system of quantification was very logical and easy therefore to actually 'quantify': A = 4 points, B = 3, C = 2, D = 1, and F = 0 points. Mediocity had been achieved, nonetheless.

So then we, as students entering high school (so grades 9–12) as freshmen, had pre-earned an academic 'value.' An equal number of As and Bs in junior high were now worth a score of 3.5. Bs and Cs would result in scores between 2.0 and 3.0, etc. Obviously, it was advantageous to have obtained all the As possible so as to post a "four-point" average or as close as possible when entering high school.

Both Margot and I had 4.0s, and we met our 'equals' at Bellevue High School in September 1956. Margot, I recall, had Jennifer Dunn and Lynn Goddess and several other 'smart girls' with which to compete. But, it was the guys' competitions that seemed to really matter to us. In retrospect, we were and behaved like classic but novice 'chauvinist' freshmen from day one. I recall John King, Victor Parker, Kenny Emanuels, but no girls, therefore, as immediate competitors. In retrospect, our now 'sexist' behaviors had begun early on!

These quantified grades, or numerical scores, became grade point averages (GPAs), which were real, comparable measures of our academic success(es) or not. From that day forward, GPAs were the measure of our success.

Mediocity had already been achieved, and complexity was well in sight. Those 'four-point' As were of absolute importance. Three-point Bs were undesirable. So we began as freshmen with six classes per day—three in the morning, then lunch, and three in the afternoon. The scholastic or 'academic year' came in three 'semesters'…actually a misnomer. There were instead actually three each, three-month trimesters: fall, winter, and spring quarters as generally recognized. We all had our summers off. There were no "summer school" options then.

So, each quarter included six classes per day times three or a total of eighteen grade scores per year. Straight As were equal to 4×18 or 72 total points. Straight Bs totaled 54, and so forth. So, with a mixture of As and Bs (Cs were unacceptable), say 63 total points, divided by 18 (classes), equaled a GPA of 3.5. Throw in some, say three Cs, and 60/18 = 3.33 GPA, etc.

This system, when first compared to the simplicity of grade school and then to the mediocity of junior high, was relatively complex. About the end of our sophomore and then the beginning of our junior years, Ms. Lear (who taught

English and Latin) and Ms. Hardy (who taught geometry and algebra) began using <u>pluses</u> and <u>minuses</u> with their letter grades. So, depending upon one's viewpoint, both an A- and a B+ might have been arguably 'worth' 3.5 points, which did not seem reasonable. So, an A- was assigned a score of 3.7, and a B+ scored 3.3. The mathematical equivalents of letter grades retained their proportionality. A flat 3.5 individual grade was essentially impossible. Complexity was in force to stay.

That's the way it remained through our senior years. JK (John King), Vic (Parker), Kenny (Emanuels), and I were "neck and neck" at graduation.

We all showed GPAs of 3.8 and above. But then, as if out of nowhere, came Richard Smaby. *Richard who?* I recall JK had a 3.93 (so one or maybe two Bs over those entire four years). I figured I would come in second with my 3.87 accum. But Smaby, whom none of us expected at all, tallied a 3.90 flat. It was a shocker for sure!

JK was our "valedictorian" (number one/highest GPA) graduate with his .03 edge over Richard. Again, I'd expected to be "salutatorian" (so number two), but Smaby aced me out by his .03 edge as well. These differences amounted not to letter grades but rather to an A+ (plus) or to an A- (minus) somewhere, sometime over our four years. I wound up number three, and there is no Greek title nor 'name' for that…just "second runner-up!" It was really that close…but a nameless 'podium' finish only for me nonetheless.

The scoring, even in high school, was again a bit complex back then in 1959, but not nearly as complex as it has become currently. Back then, again in the 1950s and 1960s (and I'd note: pre-computers, since we had none), 3.85+ GPAs were good enough for application to almost any top college or university. GPAs of 4.0+ are almost required for most of the same applications today. In fact, 5.0 GPAs, impossible then, are achievable now.

The exact abilities between top students were then decided by the SATs = Scholastic Aptitude Tests. For whatever reasons back then, maximum scores on either SAT 'half,' which were the "language or verbal" (first) half and the "math" (second) half, were 800 each—not 100 or 1000 but 800. I was best at math and science, scored well enough on the written half and 800 on the math…and was accepted at both Dartmouth College and Stanford University as a 'pre-med' major. I accepted Stanford. JK and Smaby, I recall, both scored 800s on their language exams and well enough on math. JK graduated from Harvard Law School. Smaby, as was his more 'clandestine' style, chose a more obscure university but apparently went on to obtain his PhD from MIT or its equivalent. I never knew for sure…'sneaky' as he was. He married his high school sweetheart, and the two became expert tango dancers…as confirmed via our later class reunions.

As it turned out, four of us: Vic Parker, Ken Emanuels, Margot McDonald (remember her from grade school?), and I chose Stanford. They all graduated "on-time," so in 1963. I graduated "in my time," actually twenty-one years later, so in 1984…for reasons far too complicated to fully enumerate herein.

My sister Linda, also a product of Bellevue High School, also attended Stanford. Though four years younger than I, she finished "on-time" in 1967, actually seventeen years before I did. Again, the reasons for my delay, i.e. military service and the start of a restaurant chain, were 'reasonable' interludes, but again, themselves complex and discussed no further herein!

Why all this detail? Back in 1959, again our high school graduation year, a 3.8+ GPA of any merit and a combined SAT score of 1450+ were good enough for application and acceptance to either Harvard or MIT or Dartmouth or Stanford or to almost any other top college or university. By the time my sister applied, so four years later in 1963, the GPA requirement was closer to 3.90+ and 1500+ was necessary on the SAT. This in turn required a great deal of scoring proficiency (or proficiencies) for high school grads... like plenty or better, exclusively As and A+s back then. And so going forward...

Education entered the age of "5-point" classes. These were essentially introduced in the latter 1960s about the time my sister graduated. To meet the GPA demands then of top colleges and universities, "gifted and talented" or simply "extra credit" courses of all kinds and descriptions became commonplace in high schools. It is hard to explain how an A+ grade, then the highest and worth 4.3 points, was fairly exceeded. But the "g and t" 5-pointers were awarded rather liberally...and the calculation of fair and comparable GPA scores became very, very complex indeed. And they have become only more complex today...which in turn has become a metaphor for any measure of intelligence one purports to define. Private high schools and advanced-placement studies now offer a preponderance of 5-point rather than 4-point classes. When the prior 4-point-plus maximum math no longer applies, the question becomes: Are 5-point students today any more intelligent than were 4-point students then? (Or what letter comes before A?)

Per a recent statistic: Stanford University, my alma mater, had enough quality applicants to fill its entering freshman class of 2021 (so applied in 2017) all but three times over and all with A+ or 4-point GPA or better students. The question remains: Was this class any more intelligent than was mine back in 1963?

It's hard to quantify and harder to know for sure. Will anything really change when 6-point classes become standard in advance? Complex grading systems become...what?

So simplicity in intellectual grading systems has advanced through mediocity and is now well past complexity and may itself be morphing into...irrationality? If and when presented...then how will these systems and these future students be scored?

Irrationality may become the 'stuff' of artificial intelligence. Numbers, calculations, and concepts beyond any rationality, so beyond reasonable human conceptuality, are now extant and will become increasingly so. More and more, human intelligence, so HI will be assigned to artificial intelligence, so AI as is currently demonstrated by

a.   machine learning,
b.   quantum computing, and
c.   BCI (brain-computer interfacing).

This same progression or transition from simplicity through mediocity and now through complexity has occurred in the sciences too. Euclidian geometry and Galilean math advanced through all sorts of 'mid-transition' theories and discoveries, then into particle theory, then quantum mechanics, to quantum electrodynamics, and to relativity and beyond. If the science and math of the Greeks were simplistic in retrospect, then Albert Einstein introduced advanced complexity for sure. Dr. Einstein introduced most subsequent scientists to their actual graduate schools!

And now the advanced studies and PhD opportunities have presented themselves as in 'String' and 'Gut' and

'Supersymmetry' and 'M' theories and ADD math and… and… Who can predict the next ultra-complex scientific effort(s) and achievements? Nonetheless…

It would be of interest to know somehow and in advance where the 'exact' sciences will be in more or less eighty years…say by the year 2100? Apologies to Dr. Einstein, but his complex theories have now been well-exceeded, and a new classification or description of the sciences, something like "advanced conceptuality," may apply. We may shortly ask of and rely upon artificial intelligence which, again will likely be 'in charge'…ultimately. Then what?

AI will surely far exceed any levels of human achievement as currently understood or imagined…and do so likely before my grandson (born in August of 2017) or granddaughter (born in February of 2021) graduate their universities.

They will likely experience simplicity (first) then mediocity and then complexity, then…what? What will follow complexity? "Supercomplexity"? And there should be a new word for that by then…perhaps "Quantumplexity"?

Chapter 2

Points...and the fact this author grew up on one in Medina, Washington

What is the purpose of this writing, so then what is the <u>point</u> of this chapter, specifically?

Depending upon anyone's <u>point</u> of view, one can <u>point</u> to various observations and discussions of <u>science</u> versus <u>logic</u>. Case(s) in <u>point</u> would also be the elements of <u>science</u> versus <u>religion</u>. The <u>point</u> here is then the unanswered question: "<u>Is religion logical</u>?"

It is obviously rather <u>pointless</u> to simply argue one way or the other. Rather, the <u>point</u> would best be to <u>point</u> out the <u>difference(es)</u> between science and religion. A primary <u>point</u> of contention here would be how Jesus was born to the Virgin Mary, for example...whether Joseph and Mary did or did not have sex for the birth of their son.

- Such would be a <u>pointless</u> consideration, perhaps, to an atheist, but
- It might be exactly the <u>point</u> to the Pope?

To actually demonstrate the concept of <u>points</u> more generally:

- Take a <u>pointed</u> writing instrument like a pencil or a pen.
- Make a single mark with either and with no intended dimension.
- That's a <u>point</u>, supposedly without dimension(s).
- Except that pencil or pen <u>points</u> do have them—dimensions.
- So the single, <u>pointed</u> marks they make
- are, in fact, dots, and
- dots have dimensions.
- <u>Point</u> made?

While the above may seem <u>pointless</u> to some, it is not entirely. And that is exactly the <u>point</u> of interest here, because

- <u>Points</u> must be dots with dimensions.
- In fact, depending upon one's scale of measurements…
- <u>Points</u> or dots must have both diameters and circumferences,
- Because they are actually tiny circles on paper…or upon any flat surface, which is
- Exactly the <u>two</u>-dimensional <u>point</u> here, while in fact, and further…
- They may be considered to be tiny 'flattened' spheres in space.
- The <u>point</u> being here that they truly have <u>three</u> dimensions, therefore! However,
- We've expanded upon the initial <u>point</u> of this chapter. So…

Are dimensionless <u>points</u> in space either actual or are they virtual? In concept? Or in fact?

Best case in <u>point</u> here would be "singularities," like the one that logically had to have been the <u>point</u> of our beginning…most specifically, the Big Bang, our genesis. It is thought to have been:

- a <u>point</u> of infinite density and
- a <u>point</u> of infinite temperature, and that occurred at
- some random <u>point</u> in space, and at
- some random <u>point</u> in time.

And the above are the general, accepted parameters of our universe's "Big Bang."

But the primary <u>points</u> to consider accordingly may be "Where exactly did this occur?" and "What was there, at <u>that exact point</u> prior to this occurrence?" Note: Many cosmologists and some astrophysicists have made this inquiry the primary <u>points</u> of their careers.

It is also relevant to note the following:

- Painters that paint primarily with the tips of their brushes are considered <u>pointillists</u>.
- Women (typically) who stitch pictures and patterns on cloth are practicing needle<u>point</u>.
- Men (typically) who lead out on conquests or upon projects are <u>point-men.</u>
- In the military, those in front are on <u>point</u>.
- When there is consistent disagreement, it becomes a <u>point</u> of contention.
- So then speakers and writers who demonstrate and retain consistency remain on <u>point</u>.

- Specific and primary places, objects, and amounts become reference <u>points</u>.
- Most all team sports and some individual contests are measured, and so they are won or lost by <u>points</u>…like match <u>points</u> at Wimbledon or even an extra <u>point</u> at a Super Bowl or a winning <u>point</u> at a World Cup.
- So are most more complicated card games like bridge or cribbage or casino or hearts measured in <u>points</u>, wherein the queen of spades counts as 13 <u>points</u>, etc.
- Our math, so then our metric system of measurements depends upon decimal <u>points</u>.
- Perhaps the key function of one's index finder is to <u>point</u>.
- So German Short-haired Pointers do just that…<u>point</u> out wild game birds.
- And when one exhausts most all sensible <u>points</u> of relevance, perhaps as have been herein, one may have reached an intellectual <u>point</u> of no return? Recall only that…
- I grew up in a house on Evergreen <u>Point</u> on/in Lake Washington…as this chapter's title implies!

What again has been the <u>point</u> of all the above?

- Actual (so nongrammatical) <u>points</u> (including those of pens and pencils) do have dimensions.
- And <u>points</u> with dimensions are really 'dots.'
- And dots, in reality, are like very small circles… when in two dimensions.
- But in three dimensions, dots become spheres…with or without dimensions. How possibly?

Again, the elemental structures of our universe appear to be either all very large or very small spheres, and they behave as dots, more specifically as <u>points</u> depending upon one's <u>point</u> of reference. So then...

Our entire structure and all its current contents, being the result of their original, constituent particles, were generated from an original <u>point</u> which one must assume was actually a sphere itself of some original dimension, <u>and</u> had to have been from an original source...somewhere. Again, this may be a separate <u>point</u> of consideration.

At this <u>point</u>, it may be best to <u>point</u> out specific references to chapter 16, "Herb," and chapter 17, "Dimensional Proportionality."

Lastly, and to restrict much further wordplay, the conclusions may be summarized as follows:

1. Our universe—its space, time, energy, and very existence are thought to have originated from a single <u>point</u> some 13.7 billion years ago...as our Big Bang.
2. In time, if our (or any other) universal expansion were to result in a density pressure (total mass equivalence) less than its total gravitational attraction, ours or any other universe like ours...
3. May well reverse itself and again, in time, contact to a <u>point</u> considered to be a "Big Crunch," so then the reverse of a Big Bang.

    Note again: It is the position of this writing that our Big Bang had to have been precisely that <u>point</u> of collapse of a dimensional universe that was either once ours or that existed one level up from our own. And...

4. At our Big Bang, an initial temperature in the range of at least 10^{32} degrees Kelvin was present…which temperature has been subsequently redistributed (insofar as our current microwave background temperature measures 2.7 degrees Kelvin) primarily into existing stellar objects, which have themselves coalesced into subsequent <u>points</u> of extreme temperature and pressure (read: stars) that in return have initiated nuclear synthesis.

5. Said processes of nuclear syntheses cause the release(s) of positrons, so <u>points</u> of anti-matter that annihilate with <u>points</u> of matter that are electrons and that create <u>points</u> of heat and light energy that sustain and continue our universe's existence.

6. Therefore, all heat and light, so then all electromagnetic energies expressed as photons, generate from such <u>points</u> of mutual annihilation.

7. Gravity organizes and structures our universe, and its function(s) focuses upon the center <u>points</u> of all masses. Gravity always seeks its center <u>points</u> in all masses.

8. Matter particles, both atomic and subatomic structures, and energy-force particles behave as <u>points</u> of mass and energy that interact and essentially combine to construct all the elements, isotopes, and structures with mass that exist.

9. It may prove to be that <u>point</u>-like substructures, currently identified as Calabi-Yau spaces, actually exist within any and all bits of matter, however small they may be.

10. Safe to assume also, just as photons exist as <u>points</u> of electromagnetic energy, so then do gluons and W^+ and W^- and Z particles exist as <u>points</u> of the nuclear-

force energies, so then, might gravitons ultimately prove to be the <u>point</u>-particles that explain gravity?

11. <u>Points</u> appear to be limits of our current measurement(s), regardless.

Whenever considering our existence in space and time, said <u>points</u> of reference(s) become both our starting <u>points</u> and ending <u>points</u>. The structure of all that exists theoretically came from, and so may be reduced ultimately to, a <u>point</u>.

This is with, perhaps, the exception of time. One "<u>point</u> in time" may be conceptually understandable but physically and actually impossible. Complete "stoppage of time" is conceptual only. This <u>point</u> will be discussed further herein. What, for instance, is a 'moment'? For how long does a moment exist?

However, the fact again that our very existence began at and from a <u>point</u> and that our continuation itself proceeds via <u>points</u> of stellar nucleosynthesis, which, if large enough and over time, may reduce to <u>points</u> which are black hole singularities and which may, in time, reorganize and collapse all matter that currently exists back into a massive <u>end-point</u>...if hard to imagine?

All the above certainly deserves some consideration, but what we do or do not do in the interim may not matter. Our action(s) or inaction(s) may possibly be <u>point</u>less.

Hopefully not...as <u>pointed</u> out and as follows:

So to end all prior 'word play': What <u>is</u> the proper, or better, the understood definition of a point? A German mathematician (and soldier) did compute (while on World War I military duty in Russia and apparently 'under fire') that mathematical point at which, given sufficient mass (so, matter) being drawn together simultaneously in one place, gravity would, in essence, strengthen to focus sufficiently to cause the immediate space-time to curve to the extent that light, in fact, that all radiant energies within such a space would be retained. The exact dimension of such a space surrounding this point of maximum gravity would depend upon the total mass that had accumulated and collapsed in the first place. If this is confusing...

This appears to have been also the first mathematical explanation of any Black Hole. The outer edge of such a 'black' space has been defined to be an "event horizon" of any corresponding black hole. And this calculation so impressed Albert Einstein that he incorporated Mr. Swarzchild's (the German mathematician's) calculations into his own. Insofar as...

Black Hole event horizons definitely do have varying dimensions, but their central points, currently referenced as "singularities," apparently do not? They remain unmeasured. Event horizons, as suspected, exist therefore as spheres with radii of varying lengths. But again, due to their extreme central gravitation, little (or no?) radiation escapes from within them. They are, therefore, 'unseen,' or better, 'unseeable,' hence 'black.' Again, to date, they are very difficult but not impossible to either measure and/or to define exactly.

There exist central points within all black holes that do to date defy measurement(s) at all. If so assumed then to exist without or with currently unmeasurable dimensions, they must and do exist mathematically as 'singularities.' The contradiction arises: that without (so with 0) dimension at all, their exact mathematical descriptions fail, and they become, in effect, 'virtual.' So, virtuality (or virtualities) like a singularity (or singularities) is (are) very difficult to define.

Mr. Swarzchild's calculations again demonstrated that once zeros enter any existing dimensional measurement(s), any answers, so any current mathematical descriptions of a singularity fail. It is consistent with the function of division in any equation wherein any denominator becomes zero… any 'answers' become unobtainable therefore. And this is the accepted and current 'situation' or 'disposition' of black hole singularities. However…

Mr. Swarzchild's calculations were (and are) based upon the assumed existence of <u>fixed and unmoving</u> and, therefore, <u>stationary</u> points. The facts and subsequent observations indicate that black holes are not fixed at all, but are spinning, perhaps furiously. It need be noted, therefore, that anything that spins must have an axis and equatorial circumference in <u>fact</u> to <u>exist</u>. One might say: "Nothing can't (or doesn't) spin," which, being a double-negative, means: "If spin exists, so does (some) dimension." Accordingly…

Singularities in spinning black holes must have dimensions—however small they may be. If instead, our mathematics fail, our logic must not. We are therefore presented with a conundrum, or better, a contradiction, and…

It is reasonably clear to this author that our mathematics is currently insufficient. Enter again artificial intelligence. What might it 'think'? Perhaps the next super-supercomputer will reason this out? Via a 'new' mathematics?

The point being again is that we are currently unable to figure this out. And, at what point might this change? To some conclusions…

Conclusions

1. If dimensionless, then a <u>point</u> is 'virtual' or 'conceptual' only; and it is a good definition only of 'nothing' or better, of 'nothingness.' Accordingly…
2. Then how many of such virtual <u>points</u> may exist in total is a very good definition of 'infinity.' And…

3. <u>Points</u> of 'nothing' can't and don't spin, which are double negative statements, then the following must be true:
   A. Some things can and do spin. Or…
   B. If there is spin, there is something. And…
   C. If there is something, there is dimension. Further…
   D. Given dimension, however large or however small, the total absence of anything (per chapter 15, "Absence of Anything") is doubtful.
4. It is evident that all structures, again however large or however small, are in motion, and so are moving and are with spin. Nothings (without dimension and without spin) are 'virtual' and are not 'actual,' therefore. So then…
5. Levels of structural and integral universes are logical and should be expected to exist. And…
6. If they do, then dark matter may be expected to exist as a collection of very small <u>points,</u> with spin, and, therefore, with dimension yet unmeasured. Accordingly…
7. Black hole singularities would be the most logical transition points between 'our' matter and 'dark' matter, therefore. Furthermore…
8. Dark energy must imply the existence of very large structures in at least one other much larger dimension. Therefore…
9. The unlimited expansion of our universe (and of the increasing rate of said expansion) is logical.
10. In conclusion, Size Matters…as does Dimensional Proportionality (chapter 17).

Chapter 3

CIRCLES...When Spun Become SPHERES...When Spun Become GYROS...Which Behave...Consistently

Circles were considered the "sacred solid" by most notable, early Greek mathematicians and philosophers...and for good reason(s).

Given that a line drawn upon any flat surface or in flat space is <u>one</u> dimensional and that two straight lines therein may thereupon intersect once only, then add a third line that intersects both of the first two, and any <u>two</u>-dimensional solids thus drawn or constructed similarly begin with <u>three-</u>-sided triangles. This simplest solid was first recognized by Euclid, a very notable early Greek mathematician to be sure!

Continuing with the addition of sides to solids come squares (and rectangles) with four sides, then pentagons with five, hexagons, septagons, octagons...and ultimately, circles. Circles exist as the most complex (i.e., most multi-sided) Euclidian solid of all...and again, the most "sacred," therefore.

Circles exist as two-dimensional solids with an "uncountable" number of sides. A "kilogon" would have one thousand or a "gigagon" would have a million...still not enough. Again, a circle, by definition <u>and</u> construction, exists as the most complex Euclidian solid with an <u>infinite</u> number of sides, therefore.

Euclidian geometry was first concerned with one-dimensional lines and two-dimensional solid structures. Then the geometry of three-dimensional pyramids, cubes, and spheres became obvious. Spin a line, any line, from its center and describe—or better, <u>scribe</u> a circle. In fact, spin any two-dimensional solid from its center and scribe a circle. Spin a circle from or around any diameter, or spin any three-dimensional solid in all directions from its center and scribe a <u>sphere</u>…which is a <u>three</u>-<u>dimensional</u> <u>circle</u>.

Little wonder circles were considered "sacred" early on. And again, when given motion, as in a <u>spin</u> around any diameter, they explain and behave as the primary shape and structure of most any and all objects in motion…<u>spheres</u>. And to be sure, almost everything in fact <u>is</u> in <u>motion</u>. (Refer to chapter 9, "Mario's Champagne.")

Unless fixed and nonrotational, all objects in motion tend to behave essentially as circles or spheres. Any spinning object does just that…it behaves either as:

a. a wheel, when on edge and in two dimensions, or…
b. a sphere from any central reference point and in three dimensions.

And these two observations seem to hold for both very large structures, such as celestial/universal objects, to the very small microscopic structures and down to the very, very small particle structures that make up all matter…and perhaps even energy. All atomic and subatomic particles have assigned "spins," except pions and perhaps the Higgs boson, which are uniquely separate consideration(s). Moving along…

Mathematics itself is arguably of human construct, but even in its 'numerical detachment,' if one may so articulate, the mathematic explanation of circles is quite elegant and mimics the circular parameters and multidimensional characteristics of the "sacred solids" so described and always based upon two primary measurements: 1) the diameter and 2) the radius.

1. The outer dimension or circumference of any circle is calculated from any <u>one</u> <u>diameter</u> (only), times pi (3.1416 as discussed below)…so in <u>one</u> dimension (only).
2. The inner area of a circle is calculated with any <u>half-diameter</u>, so radius <u>squared</u> (so <u>twice</u>/times itself), times pi…so in <u>two</u> dimensions.
3. The volume of a sphere is likewise and accordingly any radius <u>cubed</u> (so <u>thrice</u> times itself) times four pi over three…so in <u>three</u> dimensions.
4. The surface of a sphere comes then back to a <u>two</u>-dimensional calculation of four-times any radius <u>squared</u> times pi…notably also <u>four</u> times the area of a circle with the same radius. For mathematicians: There must exist a relationship with 4s herein that it, four, is the only number when reduced to its square root is also exactly twice that root. So $2 + 2$ and 2×2 both equal 4. TBD.
5. What then is unique and very circular in both a mathematical and metaphysical sense is that the surface of any sphere is also equal again to its <u>diameter</u> times its circumference, $d \times d\pi = d^2\pi$, which equals $4r^2\pi$…as if by circular logic!

To best appreciate the five calculations above, consider that the very unique ratio, pi = 3.1416, which explains the basic relationship of the length of any line when spun from its center (any diameter) to the length of any circle it scribes (any circumference). Pi is common to any and all circular computations…much like c^2 is to any equation between energy and mass ($E = mc^2$). And both pi and c are constants. So…

If one assigns the constant value of one (1) to be the constant value of pi, then the mathematics of circles becomes the variation of two measurements = either the diameter (d) or half-diameter (r) = the radius:

1. The circumference again is determined simply by any diameter, so $C \approx d$ in one dimension.
2. The area is determined by any half-diameter or radius squared, so $A \approx r^2$ in two dimensions.
3. The volume (of a sphere) is determined by 4/3 of any radius cubed, so $V \approx 4/3 \, r^3$ in three dimension.
4. a. The surface (of a sphere) is determined by 4 times any radius squared. So $S \approx 4 \, r^2$, which again equals…
   b. any diameter times its circumference, so $S \approx 4 \, r^2 = d^2$, both in two dimensions. And…

All these relationships, again all factored by pi, depend solely first upon the length of any diameter. Circular poetry of sorts is expressed in the mathematics of circles themselves and aligns perfectly with the three-only spatial dimensions of height and width and depth.

The computations and determinations of all circular or spherical objects come back 'full circle' to depend upon one basic measurement, the original one-dimensional line, which is the <u>diameter</u>, variously factored by 1/2, 3, or 4, and then always by pi. Circles, both by their structures and inherent mathematics, always come back to where they began—both 'sacredly' and 'elegantly' as above, and also 'physically' as below.

Again, spin a line or any two-dimensional solid from its center and scribe a circle. Spin a line in all directions or spin a circle via any diameter or spin any three-dimensional solid from its center and scribe a sphere. This geometry, given a specific motion, is logical.

The above considerations and 'spin' process appear to be most consistent with the 'behaviors' of extant cosmic, so very large universal objects in particular. Such objects as observed in space (first, presumably, via Galileo's telescopes) as circles, and subsequently, by Voyagers' (note: There were two of them) cameras, as spheres, do confirm all primary celestial objects in our solar system, so our Sun and all planets and moons do exist as rotating spheres. So, it must be permitted to assume…

Those cosmic objects that do not exist as rotating spheres (think 'tumbling' asteroids, for example) lack spherical architecture and lack spin.

Even stars, such as our Sun, and planets, such as our Earth, exist as spheres with sufficient spin and with hot and plastic interior cores, so they tend to bulge due to the centrifugal forces at their mid-lateral points (equators), therefore, at 90° away from their poles. More easily understood: Almost all celestial objects with sufficient spin and hot centers elongate at their equators, shaping themselves more specifically and technically as ellipsoids. Earth's equatorial radius is noted to be approximately 2.77 miles in excess of its polar radius, for example. It is unknown by this author how much our Sun so bulges or elongates at its equator as well…but it is understood that it does. And it is therefore assumed that all celestial, or universal, objects with hot/plastic centers, but with the unique exception of black holes, do the same…

if and when large enough and 'plastic' enough and given sufficient spin(s). Again, black holes apparently do not bulge due perhaps to their very cold centers and unique architecture. TBD.

Nonetheless, even if some are a bit 'fatter' when measured perpendicularly to their spin axes or not, all cosmic objects, including (but not necessarily be limited to) live stars or dead stars or neutron stars or black holes, and planets and moons and galaxies are all essentially <u>spherical</u>.

And, so are bubbles…for separate consideration. TBD.

Back to this central observation then: All solid and large or primarily gaseous and large universal objects, with spin, have axes. And, said axes identify equators and poles, which then demonstrate axial stability like "gyros" with properties that include

1. 'uprightness' (arbitrarily 'north' and 'south'),
2. trajectories that are more determinable,
3. orbits that are consistent and measurable,
4. consistent gravitational characteristics, and
5. other predictable 'behaviors' as well.

To <u>begin</u> with, essentially all celestial objects organize as spun circles, and to <u>end</u> with, they are all rotating spheres. But why should this matter?

Unspun circles may instead roll. Circles in their one-dimensional (simply drawn or inscribed) simplicity exist upon paper or upon graphic or projected surfaces or screens only. For a circle to actually roll, it must obtain shape and so obtain additional spatial dimensions to first become a hoop or loop or a disc…best examples being a rim or a wheel or a tire of any sort. For examples: Tip a hoop or a loop or a disc vertically, upon its edge, and roll it upon any surface, and it becomes a <u>wheel</u>. Give it continuous rolling motion via any mechanical assistance (legs or driveshafts, etc.) or some inertial momentum or simply roll it downhill, and it will align itself gravitationally and remain upright.

The gravitational alignment of a wheel upon any surface relates to its axis of spin in motion…analogous to its <u>horizontal</u> spin axis when in motion and so then <u>perpendicular</u> to the extant gravitation. Said wheels, when in motion, will remain upright and resist any force or attempt to change their vertical alignment. So, a rolling wheel (on a relatively smooth, downhill surface) will not tip over by itself. Cyclists, both human-leg-powered and motor-driven, are most specifically aware of this 'proper' or 'upright' behavior of a rolling wheel when in contact upon or across any surface. The physics is complicated, but remove or leave the surface or stop the rolling motion and said vertical stability and upright 'behavior' is lost. This, as compared to…

Spin a circle, so then actually a hoop or a loop or a disc, flatly or horizontally, and its vertical axis will instead align itself <u>parallel</u> to, not perpendicular to, the gravitational force to which it is subject…and essentially for the same reasons. Observe a gyroscope. Gyroscopes maintain their gravitational alignments whether or not in contact with any supporting surface. And this same gyroscopic behavior grants parallel or 'upright' axial stability to any and all spinning objects, so to spheres in particular…be they objects in solar systems or, presumably, in atomic structures as well.

Note: Accordingly (almost curiously?), the axes of rolling wheels remain <u>perpendicular</u> to gravitation when rolling across and in contact with any surface. The axes of gyros tend to remain <u>parallel</u> to gravitation whether in contact or not and remain relatively fixed and unmoving (i.e., do not roll). Planets in solar systems do behave as gyros, and in ours, in particular, all except one remain properly aligned, both "up" (as in north) and "down" (as in south). This is more complicated and difficult to explain because said alignments appear to be again roughly <u>perpendicular</u> to the Sun's gravitation. But...

Planets are quite specifically also in constant lateral motion, so also in orbit around the Sun. They must be subject to both circular (noticeably wheel-like) and spin (noticeably gyro-like) alignments simultaneously. All planets except number 7, Uranus, remain relatively "upright," so roughly perpendicularly "north" and "south" in respect to the Sun. Uranus, however, has been observed (again by *Voyager II*) to spin upon its "side." Its axis of rotation points roughly "east" and "west," off by nearly 90° in respect to other planets' rotations...which is an anomaly or unusual behavior and is subject to further discussion in *Fizzicks 201*, perhaps.

Over and above all: If not behaving as spinning spheres behaving as gyros in constant motion, all that we know, whether large or small, would likely 'misbehave' and fall into chaotic motion(s). Or worse, heaven forbid, if motion were ever to 'tire out' or momentum was to simply 'run out' for any reason(s), all that exists (which would likely be unknowable to us) would be utterly disorganized and chaotic and uninteresting. Such lack of organization is essentially unthinkable, so <u>totally</u> <u>uninteresting</u>. Spin is critical.

Chaos is disorganized and is to be avoided as <u>un</u>interesting. Organization, given understandable motion, therefore, is interesting.

And, the relationships and behaviors we observe as evidenced in the physical world <u>are</u> <u>indeed</u> <u>interesting</u>, <u>totally</u> <u>interesting</u>! And so many are so evidenced in life's experiences, and so then in <u>our</u> <u>meta</u>physical world as well. Concepts and behaviors expressed circularly are seen and experienced, also sought out purposely in biological, psychological, and social circumstances as well. "Cells" of most all kinds and in at least one aspect are round, for example. Such is comparably demonstrated as one may observe or experience in:

- time, just <u>rolling</u> along,
- <u>rolling</u> stones gathering no moss,
- <u>circles</u> of friends and associates, and…
- <u>circular</u> reasoning, discussions, and agreements…or otherwise <u>circular</u> <u>dis</u>agreements.

Both seem to provide a sort of 'axial,' so understandable and predictable or

- human architecture and inherent <u>circular</u> 'behavior,' which is again demonstrated as is expressed in…
- <u>circles</u> of friends and/or of opponents, that when extended broadly and/or in multiple aspects or directions (so <u>spun</u>), become…
- <u>spheres</u> of influence…spun circles again!

And then, more obviously do exist:

- <u>circular</u> exchange with <u>circular</u> coins that are <u>circulated</u>. (Paper currency is nonmetaphysical, perhaps?)
- <u>circular</u> clocks and watches (specifically <u>excluding</u> digital ones, which are decidedly <u>nonmetaphysical</u>!).
- <u>circular</u> compasses (See note A, just below.)
- <u>circular</u> or <u>tubular</u> pipes and veins, rockets, and projectiles (See note B, just below.)

- spherical objects that are "balls" in sports (See note C, just below.) Even…
- fluid dynamics expressed as spiral weather (See notes D and E, just below.)
- spherical bubbles. TBD.

And then conceptually in the sense of time and place:

- hours, minutes, and seconds of time are in measured parts of circular clocks.
- degrees, minutes, and seconds of direction are also measured in parts of circular compasses.
- day and night (subject to Earth's spin rotation).
- seasons (subject to Earth's roughly circular orbit and spherical axial but variable alignments).
- years (subject to Earth's circular, more specifically to its elliptical orbit), and then…
- universal sense of "up and down" and "right and left" (due to Earth's spin axis) and the gravitation of the Sun, which behaves spherically and spins as does Earth.

So much of how we, *Homo sapiens*, behave and manage our behaviors, relationships, motions and emotions, and then also measure our time and find our direction(s), again be they north or south, right or left, or up and down, are based upon Earth's circular orientation and its spin.

Note A (of just above): Compasses. One is a very simple device or, better, a simple drawing tool that pivots upon a center point of rotation with an angular attachment to a marking device that will define a radial dimension so one-half a diameter, and then, when spun, scribe a circle. The other is a more technical, circular device or tool, with a circumference divided into increments or degrees of arc (not of temperature) that will define the magnetic alignment of a rotating needle to determine direction. And both are referred to as "compasses," curiously.

Note B (of just above): <u>Circular or tubular objects</u>. While stationary objects such as buildings tend to be angular, more specifically square or rectangular, rockets and various projectiles (so nonstationary objects) are round or tubular or conical. Most are actually extended circles or rods, usually tapered in the direction of travel, so then are elongated cones. Circular, not angular, shapes correspond to objects with motion. Bullets in particular are given very specific 'spins' via barrel 'rifling' so to organize and improve their trajectories when in motion 'ballistically.' Circular, not angular, shapes correspond also to objects and devices that transport fluids. Pipes and tubes, veins and arteries logically comply. Few, if any, triangular or square pipes nor projectiles exist. Endless examples <u>do</u> exist. Of note also: The "Borg's" cubical spaceships (aka *Star Trek*) made no sense at all! Such is good science <u>fiction</u>!

Note C: Most all sports that are more specifically 'games,' not simply races or other timed events, utilize various modified spheres, which are <u>balls</u>. It would take too much space and time to list them all. Most balls are spun in their process of 'play.' Structural exceptions of note would be American football or rugby balls, which are elongated and so pointed on both ends. Said 'pointy' balls still orient and stabilize themselves when spun, so when passed or lateraled…for the same reasons and purpose as bullets are spun. Endless examples apply here as well.

Note D: Weather (the atmosphere is a fluid) and so behaves in rotational cycles. Extreme weather events, in particular, organize themselves circularly. Hurricanes and tornados are self-explanatory, always spun. Hurricanes in particular are large enough, so they involve enough atmospheric mass that they are subject also to the coriolis effect. As the Earth rotates to the east, any object, so an <u>atmospheric molecule</u> as affected in this instance, will travel toward the east fastest at the equator. Moving north, objects, so molecules of air included, travel a bit slower and would be ultimately and relatively nonmoving. So they spin only as affected in unison by Earth's rotation at the exact north (and same for the south) pole(s). Viewed from an equatorial aspect, the resulting atmospheric rotations occur counterclockwise north of the equator and clockwise to the south. Tornados (and water that drains in sinks and toilets) spin also, but for more specific and different reasons that may or may not relate to the coriolis effect. It is a varied symphony of the <u>circular rotation</u> of fluids driven by the Earth's 'easterly' rotation, nonetheless.

Note E: Sinks and drains (again), curly hair, pigs' tails, and the spiral rotation of seeds in sunflower blooms et al... all circular or spiral. None above seem to possess sufficient mass(es) to be affected (like hurricanes) by the coriolis effect...for the record and for those who might inquire! Whirlpools? Not sure! Refer to Davie Jones. Ocean currents are driven primarily by other fluid dynamics. But any natural motions at all express circular curvature in whatever form or via whatever curvatures they choose...all due to the similar geometry of the Earth, its rotation and gravitation and orbit... and the 'sacred' geometry. Motion generally abhors straight lines and avoids abrupt corners.

Most all considerations of cosmic objects, again existing as rotating spheres, and so then also as three-dimensional 'gyros,' apply. And essentially the same observations <u>seem</u> to apply when considering atomic (and presumably subatomic) particles as well. Per Dr. Sagan's original reply to my first inquiry, seemingly <u>regardless</u> of the specific energetic forces of <u>attraction</u> that are involved, both cosmic and atomic organizations appear to be based upon spinning spherical orientations, which respond to either:

1. gravitational force that exhibits attraction <u>only</u> for large (so massive) cosmic objects that exist relatively far apart and that are not affected by…
2. electromagnetic forces that can exhibit <u>both</u> attraction and/or repulsion or expulsion <u>only</u> for very small atomic (and subatomic) particles that are very close together, and are thought not to respond directly to gravity, due to their absence, or near absence, of mass.

Dr. Sagan's reply was and remains quite specific and points out the <u>energetic</u> <u>differences</u> <u>in</u> <u>attraction.</u> Only those that apply to and for the two systems: one very large and distant (that depends upon gravity) versus one very small and intimate (that depends upon electromagnetism) are involved. However, their very similar geometric organizations remain uncontested, perhaps ignored…this as best can be ascertained.

Going back in effect fifty-plus years, had I the opportunity, I would reply to Dr. Sagan that I <u>have</u> since, indeed, "studied more physics" and "cannot yet assign the similarity of cosmic or solar-system organization and its geometry with the same organization and geometry of atoms…solely to <u>coincidence</u>."

First of note would be: Both cosmic organization of solar systems (and of galaxies) and particle organization of atoms (and of elements <u>and</u> of compounds) depend solely upon <u>attraction</u>. Again, said organizations depend upon said attraction, either <u>gravitational</u> as between planets and stars or <u>electromagnetic</u> as between electrons and atomic nuclei (protons therein).

It is of note that little or no use of the <u>repulsive</u> forces in the electromagnetic world seems to apply to basic atomic geometries. Repulsive forces do apply within chemical reactions and the relationships between any two like-charged electrons and like-charged nuclear particles (again, protons) and presumably their like-charged contents = quarks (refer to chapter 12 "Particle Gender"). But, said repulsion, for whatever reason(s), are dealt with by nuclear forces and primarily by gluons, specifically subatomic and nuclear particles that have little or no apparent effect upon the basic geometry of an atom itself...a subject worth further discussion. TBD.

Again, both systems, so both geometries of organization, be they either very large and massive and distant or be they very small and of minimal mass(es) and intimate, both systems apparently involve spheres with spin and so demonstrate orbits that demand attraction. While the orbits may differ in mechanics, the spin requirement appears to be consistent. Essentially, all cosmic objects in our solar system rotate or spin...as it may be assumed all others in other systems do as well. An electron (in fact, all fermions that exist as 'matter' particles with mass) is (are) assigned spin of one half (1/2). Particle spins relate to a very complicated Plank constant and quantum mechanics...and may be regarded as "up" or "down" spin(s)...but remain <u>spin(s)</u>, nonetheless. Also for further discussion.

All baryons (which are force, not matter, particles and are typically assigned minimal or zero masses) are assigned spin(s) of one (1). The graviton, when isolated and identified, is predicted to exhibit spin two, however possible, and again for further discussion, perhaps. Particle spin assignments are more esoteric (currently) and, like atomic weights, are arbitrary…so they must be assumed more specifically as analogous only to the easily identified spin(s) of cosmic objects. But, again…

Spin must and does remain as spin. And, the behaviors of spun spheres, so of gyros, must apply. The Earth, for example, displays a spin of "one day," as in a twenty-four-hour rotation currently. The Sun displays rotation of twenty-five to twenty-seven days at its surface upon its equator, but slower and so longer (±38 days) near its poles. Given the Sun's contents, being superheated gases and plasma, under extreme conditions, such varied structural and internal spin rates are understandable. Jupiter, a gas giant and almost a sun itself, spins the fastest in our solar system. Uranus is again mis-aligned by almost 90° and spins on its side. Regardless, all planets demonstrate spin and exist as rotating spheres and essentially as gyros, nonetheless.

And, it cannot be simply coincidental that suns and planets do organize themselves so much the same as atomic nuclei and electrons do…not logically. Because…

Spinning circles are again <u>spheres that behave as gyros</u> and express proper physical orientation(s) due to their (however small) gravitational (not necessarily just electromagnetic) orientations. The actual effects of gravity may be infinitesimally small on particles and absent upon photons, gluons, and gravitons, which are massless. But again, the similar organizations of fermions (so electrons), so of lighter gyros orbiting more massive, rotating nuclei (effectively, collections of neutrons and protons), this as compared to relatively lighter, rotating planets orbiting more massive, also rotating stars…these relationships cannot properly be put off entirely to "coincidence," can they?

Apologies then to Dr. Sagan, since departed, but again, why does this matter?

All the above exist as primary systems that require attraction as the centripetal force to offset the centrifugal force generated by obviously nonlinear orbits. Without either gravitational or magnetic attraction to keep these systems organized (and so the occupants and various elements in proper orbit), planets in solar systems and solar systems in galaxies, and/or electrons in atoms and atoms in elements and compounds would spin away in utter chaos. And such again, to avoid chaos, requires attraction. And, this requirement applies regardless of the two, different energetic forces that control and that provide said attraction. Attraction is attraction either way…be it gravitational or magnetic. And all of the organizations known to exist are in motion and basically in a circular or spherical orientation as a result.

However many times it must be repeated: It is a very consistent 'picture' of orbiting spheres that behave as gyros and that remain in either circular or spherical or ellipsoidal or even spiral orientations…all being consistently circular and dependent upon attraction. Whether considering solar systems or galaxies, atoms or elements, centripetal force (again either gravitational or electromagnetic) must equate with the centrifugal (inertial) force that applies in/upon the basic geometries. This would seem to be mandatory…or chaos would surely prevail otherwise.

"Valence ring" requirements, the assigned orbits for electrons and the specific 'requirements' for the number(s) of said electrons in each ring, apply according to a very sophisticated set of quantum mechanical rules on the atomic scale for sure…and are beyond the scope of *Fizzicks 101*. Said rings that are laminate (like 'bent' plywood or onion rings, for examples) and that are spherical might permit intersection and collisions between electrons if not for their electromagnetic repulsion. Said rings may compress or may contract as lighter elements combine to form more massive elements. All this electron 'behavior' is electromagnetic and is again subject to specific rules...but does not prohibit the geometry. As said...TBD.

A planet's orbit around a star is also laminate or nearly laminate but is also instead planar. Planetary orbits allow only one occupant, so keeping proper distance and single occupancy also avoids collisions, without repulsion. And this slightly altered orbital architecture still permits the same basic geometry, which again is consistently circular, however observed, in one, two, or three dimensions.

Summary number one: Either form (circular geometry) must follow function (attraction), or function must follow form. Which controls? Some deeper logic must be involved.

Both the above cosmic and atomic, so then (assumed) any and all systems and/or physical organizations must be circular...or perhaps elliptical depending upon masses and initial orientations and orbital speeds. Further...

All the above again assumes any and all participants must be spherical...or elliptical, depending upon spin rates. And...

No other, 'noncircular' options seem to exist. The most symmetrical and perfectly spherical object in existence at all must be a solo, but again geometrically perfect <u>black</u> <u>hole</u>... as will be discussed in *Fizzicks 201*. But...

General orientations of multiple participants in any 'shared' system seem to (have to?) relate gyroscopically. And...

<u>Wher</u>ever or <u>why</u> ever, let alone <u>how</u>ever else, might any <u>other</u> system or organization exist at all? Seemingly

and logically, any other organization cannot…and does not! Witness again the 'sacred' stata of circles.

So, these questions may be asked of the departed Dr. Sagan: "Do the two very different forces of attraction, again one being gravitational for large objects versus the other being electromagnetic for small objects, demand the specific celestial and atomic geometries that obviously <u>do</u> exist?" Or…

"Do the masses of the participants, both their sizes and distances of separation in general, demand instead two different kinds of energetic attraction(s) to keep them in order?"

<u>Which</u>, initially at our genesis, the Big Bang, <u>was the requirement</u>?

1. Both gravity and electromagnetism to keep the two different systems organized. (And this might be assumed illogical because 'who' was to 'know,' early on, that a universe organized gravitationally only would result?) If so, then when/how and/or in what forms did gravity and electromagnetism even exist prior to recombination? And for what purposes? Or…

2. Given the circular/spherical organizations required, let alone available for any and all purposes, then actually three very different kinds of attraction and containments would be necessary again:

   a. electromagnetic for very small objects with negligible masses and very close together, just as our universe existed for its first ±380,000 years. Or…

   b. gravitational for very large objects very far apart, which have presented themselves since?

    c.   nuclear containments…discussed separately.

Which came first: the geometric and organizational requirement (only) or the force requirements (both and all)? Which came first: form or function? Did the geometry require the attraction and containment, or did the forces involved create the geometry?

Summary number two. Any and all systems and general organizations extant and of whatever sizes and therefore of whatever masses, <u>and</u> subject to whatever distances of separation are again of essentially the same circular geometry. And, any and all of the participants or entities or objects, be they stars and planets or (primarily) protons and electrons, seem to exist as rotating spheres.

Two versions of circular, orbital organizations exist regardless:

1.   <u>Planar</u>, as-in 'wheel-like.'
    a.   <u>Planetary</u> orbits around most all stars, being confirmed, and…
    b.   Lunar orbits or debris (like either Earth's moon or Saturn's rings or both) around almost all planets.
2.   <u>Spherical</u>. Same as for…
    a.   Electrons. Same and…
    b.   Subatomic particles such as the following:
        i.   Gluons around any nucleons, and…
        ii.   Gluons (also) to keep protons in any nucleus from flying apart, and…
        iii. w+, w- and z° particles (also?) around any nucleus to control radioactive decay.

Any and all subatomic (gluons' and others' too) functions appear to involve spinning spheres as participants that cause resulting organizations to behave circularly due to common <u>attractions,</u> or better via necessary <u>containments</u>. And this is expressed regardless of either the source or reason or requirement for either attraction or containment. Gluons, even of zero mass, retain quarks of very small masses in spherical organization, as do W and Z bosons, with masses (a separate discussion) that keep generally circular atomic nuclei behaving and functioning 'properly' and 'predictably.' The mathematics of said particle physics is complicated, but the logic and geometry are not. TBD.

Apparent in any system: As orbital alignments and angular momentum rates increase for planets, then said orbits may express various celestial alignments that result in elliptical (so then not exactly circular) orbits. So then also as any planet's (so any participant's?) spin rates increase, they (celestial) objects may, but it is unknown if atomic or subatomic particles do, themselves, bulge at their equators and also reshape to become more elliptical.

It is additionally apparent that galaxies organize themselves similarly, but in spiral orientations that result in lens-shaped geometries when viewed laterally. So they organize somewhere in between planar (wheel-like) and spherical (atom-like) systems but, on the 'grand scale,' they seem to be essentially <u>coriolinear</u> and/or <u>circular</u>, nonetheless.

Whatever else may be said or assumed or imagined about the (now) four (so to include the two nuclear strong and nuclear weak) forces that apply <u>containment</u>, it is evident that <u>they all</u> are, so then <u>it</u> is <u>all</u> first and primarily organized "<u>in the round.</u>" Therefore...

It seems most logical that <u>form,</u> so the <u>geometry first</u> applies, insofar as both our universal, atomic, and/

or subatomic organizations must and do exist with circular organization(s). Again (and again), without such circular geometric organization all subject to attraction, however provided, then chaos would result at and upon any level. And, if such is true, then...

The energetic forces (so functions) that are involved (and must have been in the more specific case of gravity) appear to be <u>secondary</u> <u>requirements</u> to the <u>primary</u> <u>geometry</u>.

Therefore...observe the "sacred solid" which is the apparent basis of all organization and of all geometries that apply and control...and which is evidenced as the <u>spun circle</u>.

No wonder the Greeks had such appreciation so long ago...as should we all! Circles are indeed <u>the</u> "sacred solid," and their motions, expressed as in their unique spins, determine the general organization(s) of their also (always) spherical geometrics...and are subject also (always) to attraction, however provided.

Damn, I wish Dr. Sagan were alive to respond! Atoms and solar systems must necessarily be organized similarly... and not just by coincidence.

Chapter 4

Day One (and the Dawn of Intelligence)

Condolence is offered in advance to anyone who embraces either the concept of "flat earth" or the belief in a very abbreviated "6,000+ year existence" of our planet. Neither apply. Our planet, which has been confirmed to be roughly 4.5 billion (with a b) years old, is a sphere and is, importantly, "the third rock" from our Sun. No exceptions nor denials apply.

A unique set of circumstances, in fact very unique, exists that our planet, Earth, is (and/or has the following):

1. A rock. For if it were of some other material, like a gas, as Jupiter and Saturn appear to be, life as we explain and experience it would not exist. Rock makes a good, livable planet, so something to stand upon, and where wheels (and feet) work, etc.

2. The "third" in orbit…insofar as our planet is neither too close nor too far from its star, which again is our Sun. If too close, it's too hot. Water boils and any 'climate' would be intolerable. Too far, and it's simply too cold. Everything 'freezes.' And so then Earth is…

3. Wet...that is, covered quite liberally with liquid water. With<u>out</u> liquid water, life again as we explain and experience it would not exist. In fact, it is doubtful that life could have begun at all. And Earth has also...

4. An atmosphere...that is breathable. Planet Mercury (rock number 1) has none due to its small size (so low gravity) and proximity to the Sun. Venus (rock number 2) has one, but is primarily a mixture of very hot CO_2 and sulfuric acid, which is <u>not</u> breathable. Mars (rock number 4) has, or better, <u>had</u> one which has since almost entirely 'drifted,' or better, 'evaporated' off into space. And our rock number 3 has also a great variety of...

5. Elements...most specifically those which have combined into...

6. Organic compounds. And organic compounds are clearly requisites for the existence of our <u>carbon-based</u> life as we explain and experience it. Further, Earth is...

7. The 'right' size. Any circumference much smaller (Earth's equator measures ±24,874 miles) and our rock would have insufficient gravity...resulting in numerous problems. Any circumference much larger, and it is doubtful life would have progressed much beyond bacteria, maybe earthworms or possibly land crabs...maybe! And Earth has had also...

8. Plenty of history. Specifically, plenty of time equal to roughly 3.5 billion (with a <u>b</u> again) years for life thereupon to have begun and evolved.

The above eight general circumstances and literally hundreds of sub-circumstances have occurred and have combined to permit us all. We are one of millions of (remaining…having survived numerous extinctions) living organisms…up to and including ourselves, currently, *Homo sapiens*. And we occupy, if Earth's history were compressed into one year, about the last few minutes thereof. Not much time, again given our planet's history. And…

What indeed and truly identifies us, *Homo sapiens*, as apart from any and all the other life forms?

<u>Intelligence</u> does. All the rest of any species or sub-species have it too…including bacteria and sponges and trees and termites. But to date, we, *Homo sapiens*, have the most… <u>intelligence</u>.

The first, so original occurrence of life on Earth is a convenient metaphor to the first, so original occurrence of <u>observable</u> <u>intelligence</u> itself. Taking a step back, intelligence some currently <u>un</u>observable sort, call it "original," has been operative for as long as time itself. How then do two hydrogens 'know' how to combine with one oxygen to create water? What is any other 'reason' that they, Hs and Os, do this? How about two oxygens that combine with one carbon to make/create CO_2? Same questions…why? And, how? Or farther back to the Big Bang, what kind of intelligence was responsible for "recombination," for example? One may ask the same question(s) that apply to the many phases of our early universe. There must be an original plan…an original set of rules.

And this, better, <u>these</u> processes above and beyond, so millions of them, proceed with a relatively small list of original particles (fermions and bosons…discussed later) and which utilize a relatively small list (ninety-two currently, naturally occurring) of elements, that have produced millions, or better, billions of compounds or in(non)organic substances…always in the same, deliberate, and exacting proportions and via the same, exact processes <u>every</u> <u>time</u>. All these particles and elements and compounds and substances that result 'behave' the same and have done so exactly since the beginning of our time.

To say that this is "simply the way things happen" is intellectually insufficient. So that to say that our beginning, so our genesis, the Big Bang, was "just a point of infinite pressure and temperature" that just "happened at some random place in space" is also insufficient. We actually know approximately <u>when</u> it must have happened and have estimated its initial temperature, but "how" and "why" and "where" it happened and its actual source remain unknown and unexplained to date.

There must be an understanding, so an organization of "why" and "how" this all as in all that exists in our universe, "behaves" in the same way(s), admitting that we may 'know' <u>most,</u> but certainly not <u>all</u> of them? To fall back into thinking that it is simply random, or all coincidental, or all...whatever...whatever is "<u>un</u>knowable"...and is again intellectually <u>un</u>acceptable.

The "whatevers" above will, for now, be considered as OI = <u>Original</u> <u>Intelligence</u>. It (they) might also be considered collectively as <u>universal</u> intelligence, but again, owing to the possibilities of multiple universes, we'll stay specifically with <u>our</u> OI = <u>Original</u> <u>Intelligence</u> acronym.

So then, now we may attempt to define <u>Day</u> <u>One</u> of LI = <u>Living</u> <u>Intelligence</u>. Science and cosmology have it that a tepid pool of water, water rich in dissolved chemicals, <u>in</u>organic at the time, but now since reorganized and then considered organic, was perhaps struck with lightning, with just enough electromagnetic energy to either form (note 1) or to activate (note 2) a complex molecule, call it "primal DNA" or "primordial DNA," wherein this molecule was either instructed (note 3) or empowered (note 4) to be able to reproduce itself.

It was just both that simple and that complicated...at that moment. An inorganic molecule became organic and so became able to reproduce itself. It converted or somehow incorporated OI to be exhibited as LI...a <u>L</u>iving <u>I</u>ntelligence. And LI has since evolved into HI...<u>H</u>uman <u>I</u>ntelligence... currently our <u>own</u>.

This process of evolution of intelligence has been functioning here on Earth for some, estimated, 3+ billion years. Without further specifics, it is estimated that that original event or the specific 'bolt of lightning' struck and formed or instructed that molecule of primordial DNA to reproduce...±3.5 billion years ago. All arguments and alternative theories aside, said transfer of OI into LI happened a long, very long, time ago, which may be analogous to the beginning of <u>preschool</u> (chapter 1, "Simplicity—Mediocity— Complexity") on Earth. And...

If that event <u>did</u> begin Earth's preschool, it lasted an uncharacteristically <u>very</u> long time, essentially almost 3.0 billion years until the <u>Pre-Cambrian</u> Age, which may be analogous to Earth's <u>grade</u> <u>school</u>...a relatively short period of Earth's history. And so to follow, the <u>Cambrian</u> <u>Age</u> itself would be analogous to Earth's <u>high</u> <u>school</u>...with a great deal of organic expression, time and assumptions included.

If then there is merit in even the <u>sequence</u> of the above, it becomes a very rich source of discussions of when, and so specifically at what time Earth's <u>college</u> <u>education</u>, first <u>undergraduate</u> and then <u>graduate</u> school, began. And therefore, if *Homo sapiens* has, in fact, appeared from <u>graduate</u> <u>school</u> with any sort of <u>master's</u> <u>degree(s)</u>, what is better? What <u>lies</u> ahead?

Intelligence has proceeded from simplicity through mediocity and well into complexity. So what is next to occur in this/its progression? What will be synonymous with *Homo sapiens'* master's degree(s) and, ultimately, its PhD?

It is assumed and will be presented herein that we, *Homo sapiens*, have, <u>with specifically the introduction of computers, digitized information and the internet</u>, we have begun our PhD studies. But...

If we continue using exclusively the nonhuman tools available to us, which tools again are computers (so machines) and digital (so nonhuman) information accumulation via the internet's (so again, a nonhuman infrastructure) brain, then what have we done? What have we created?

The answer is simple and obvious: We have created AI... <u>A</u>rtificial <u>I</u>ntelligence. And the difference herein <u>is</u> unique. To date, more or less over the last sixty to seventy years, which is the incredibly short actual lifespan of AI to date, we have, better organic life forms all prior to and including us have, so we <u>all</u> have utilized our own organic brains exclusively, this <u>without any outside help to date</u>. But now we choose...

AI, which is and now exists as our <u>outside help</u>. And what is now so intellectually and potentially so self-destructive is the fact that AI, our current outside help, <u>is</u> much more capable and <u>is</u> therefore, or <u>will be</u> therefore, much smarter and/or better educated than are we/ourselves. In time, there will surely be "no-contest." AI has capabilities much beyond and much advanced and faster...and has capacity well in excess of our own. Our PhDs will not be our own!

Evolution of intelligence is essentially the same as the evolution of life. And, as long as it remains <u>organic</u>, so then it remains a vestige of its original "self." We, *Homo sapiens*, can control it as long as it remains <u>as</u> our version of "ourselves"...as HI. But again...

46

As we (increasingly) assign it to nonhuman <u>inorganic</u> brains and to nonhuman <u>inorganic</u> processes, as in computers, digitized information, and the internet, then at some time it, so AI, will evolve and replace us and so will replace our HI. There does not seem to be any possibility of any other rational conclusion. AI will ultimately replace HI.

When this will occur is currently subject to some debate. It is notable that Mr. Elon Musk, owner and originator of notable HI enterprises currently, such as Tesla, Space X, and others, recently stated in specific regard to AI: "We (*Homo sapiens*) have contacted the demon (AI)." Well said. Mr. Musk has great insight. However...

He has also recently begun his new enterprise, Neuralink, with the express intent of connecting the human brain (again, our HI) directly to the internet (so directly to AI). This process will likely be done with the insertion of (inorganic) smart "chips," miniature smart phones, connected into our (very organic to date) brains. It's difficult to imagine the results therefore and therefrom. If we are currently concerned with the unequal distribution of wealth and/or of education, or of resources, or opportunities, or of whatever between members of our species now, one can only imagine this same concern if it will then exist (?), between the "<u>chip'd</u>' and the "<u>un-chip'd</u>" at some future time. Mr. Musk's demon is real. And so...

Suspicions of Mr. Musk arise to the end that he must be very hypocritical? One may, so <u>he</u> may, "contact the Demon" as long as there is sufficient demand and profit in doing so... apparently?

That appears to be the current condition of mankind, so then of our (current) species. Our lust for intelligence (and more specifically for the wealth generated therefrom) somehow must, and obviously will, proceed unabated via the much more capable <u>artificial</u> <u>intelligence</u>. Again, in time, there appears to be no contest! Again, AI will have capabilities well in excess of our own, and at some point will likely recreate or manufacture itself.

It follows that our <u>lust</u> trumps any of our concerns, however. It appears to be a very exciting voyage, effectively just another collective enterprise in itself. Not much, if anything, appears possible to restrain, let alone to contain nor to direct, let alone to redirect, its current disembarkation, so to speak. AI's proverbial <u>ship</u> <u>has</u> <u>set</u> <u>sail</u>.

We, *Homo sapiens*, have committed ourselves to the transition from HI to AI. Some predict that the occurrence will indeed happen sometime prior to 2100, noting that is currently a very short seventy-plus years hence. So it may instead take one hundred years to the "singularity," (which is not the center of a (to-date) theorized black hole). So the "singularity" of reference here is the actual union of man (so HI) with machine (so AI)…specifically, when AI's intelligence equals then incorporates <u>its own</u> with <u>our own</u>.

Said singularity will not necessarily be evidenced by the creation let alone the manufacture of some sort of super-robotic human, like a Marvel character, an advanced "Iron Man," or Mr. Schwarzenegger's alter-egalitarian "Terminator." It is instead the expected <u>merger</u> of man and machine, insofar as machines will incorporate human <u>consciousness</u> rather than somehow create their own. Or perhaps viewed in reverse, human bodies, all body parts themselves, except perhaps actual brains and organic eyes (both may be better shared than recreated or manufactured), will be replaced. Rather than eat and drink and rather than reproduce in our delightfully 'old-fashioned' way(s), super or semi-robotic humanoids will likely plug in and recharge (like electric cars can and do now) and go to a 'shop' or to a laboratory, not to a hospital or midwife to repair or to reproduce…but to <u>be</u> repaired or replaced instead?

This is futuristic, as intended, and verging upon science fiction. Apologies here, but it is most difficult <u>not</u> to see a time when AI first merges with and then replaces HI. And from there…

AI is most likely to morph into EI…<u>E</u>xistential <u>I</u>ntelligence…so then into (back into?) a totally <u>non</u>organic form. If and when this were to happen, *Homo sapiens* will not! Hence the title "existential." It, AI, seems to be the only real, <u>existential</u> threat to *Homo sapiens* therefore. Short of a cosmic collision with a very large, currently unknown, and unexpected asteroid or a nearby 'supernoval' event, not much else is capable of replacing us all.

Climate change will likely cause irreparable damage and dislocation…and yes, some death…but is survivable.

Nuclear catastrophe, as in a war between 'superpowers,' could come close, but again, somehow and someplace, like Lapland or upon some remote island, remoteness and shelter will provide survivability, therefore. Or…

A super virus, as in a form of Ebola or even common flu or super coronavirus that is unstoppable, might eliminate most but not all of us. Some of us may either be immune or somehow uninfected…perhaps (intentionally?) in a space vehicle or sterile facility, somewhere? It is easy to speculate… impossible to know for sure.

It is again <u>far</u> too easy to continue to theorize (perhaps fantasize?) here, but AI will surely morph into EI at some time. Give it a very short one hundred or longer two hundred or eight hundred or even one thousand years, it is now and will be ultimately "no contest." Intelligence (EI) seems unavoidable.

Will our EI be alone? It is easy to imagine another sun and its solar system and so then another inhabitable planet… the latter that may have originated 4.6 billion, not just 4.5 billion, years ago, as did ours. In fact, the mathematics, better the logical probabilities, of this having (already) occurred are overwhelming. But given the (relatively slow in this instance) speed of light, so then the same speed of information (accordingly of intelligence therefore) we, either our HI, and/ or then our AI and possibly even its ultimate expression, EI, might be unaware? The news of 'them,' or better of 'it,' has not reached us yet? Nonetheless…

Such a life-sustaining planet, 4.6 billion not just 4.5 billion years old, would present an original civilization 100,000,000 (that is one hundred <u>million</u>) years prior, so older and assumed more advanced than our own! One can only imagine the possible technology of <u>our</u> civilization, were it possibly able to continue to the year 100,002,023… however <u>un</u>likely for us!

Our Sun might well have dimmed or changed, so we, our form of our EI, would have, or better would <u>have had to</u> either move itself <u>or</u> have moved our planet to another acceptable orbit or acceptable star system. All sorts of futuristic scenarios come to mind. But one thing would seem to be assured, that…

Our EI in whatever form it may have taken, or may have then assigned itself, would likely <u>not</u> be alone, but instead would have communicated with, if not effectively merged with other then extant EIs into CI, <u>c</u>osmic <u>i</u>ntelligence. And then, if at or in such an unimaginable time or circumstance, would not CI, cosmic intelligence, have to be much the same as OI, original intelligence, was and is?

Would not the proverbial "circle," as diagrammed below, become complete?

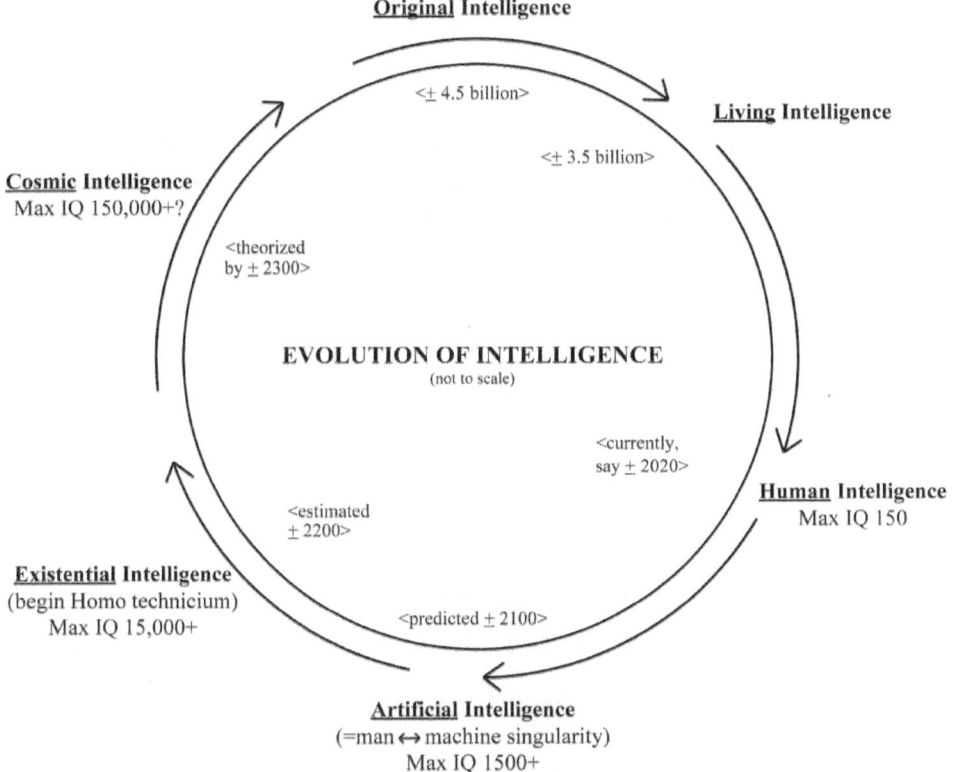

The proposal herein assumes the evolution of intelligence:

1. From OI to LI via the primal pond, the proper chemical structures, and just enough energy…a long time ago.

2. From LI to HI, which obviously <u>has</u> evolved and so <u>has</u> occurred and <u>is</u> demonstrated currently.

3. From HI to AI, which obviously is in process. All sorts of debate may apply, but then, given time, AI is sure to replace *Homo sapiens*. As in…

4. From AI to EI, which is, yes, is <u>arbitrarily</u> <u>existential</u>. Note that: We may well exist during this transition

as *Homo sapiens* that might co-exist as a subspecies of *Homo technicium*, if at all. Then…

5. From EI back into CI, which is purely theoretical. But it is logical to assume that any other civilizations will most likely have to have experienced transitions #1 through #4 above. And if so, assuming their connection(s), some form of communication, be it Luke Skywalker's recent (okay, fantasized) version of astral projection of the "force," or whatever, CI will and must be ever so similar to (so possibly the same as) OI? And the above circle will complete… as in…

6. CI = OI

The original intelligence of our universe may be difficult to define, but is impossible to deny. Some form(s) of original organization(s) and 'rules' do apply, such as:

a. light's speed limit, which is exactly the same as…
b. gravity's speed limit, which is absolute and may well be the same as…
c. the speed of time…or
d. electrons that weigh exactly $1/1836.1^{th}$ of a proton, which weighs…
e. 2eV less than a neutron because…
f. neutrons have one more 'down quark' than do protons, and…
g. 2 only electrons are 'permitted' in atomic ring number one, and…
h. 8 only permitted in rings two and three, and…
i. 18 per-ring(s) permitted thereafter. But why? And…

j. One might go on almost endlessly as to how many absolute rules and behaviors must apply… <u>always</u>. They must exist collectively as Original Intelligence.

All of the above must be of intelligent design of some sort…yes? How have these original 'rules' been set?

It cannot possibly be left to "happenstance" or "coincidence" nor to some random "occurrences" or set of just random "rules," which have been mysteriously assigned or assumed by whomever or whatever. Currently, these questions and considerations may define the extant boundaries between science and religion? Recall only…

At one time, we, *Homo sapiens*, worshipped the Moon. And it is a given that AI and/or *Homo technicium* <u>will not</u>. So when *Homo sapiens* is replaced, consideration of 'creation,' as currently understood, might become redundant? We, *Homo sapiens*, may be very 'temporal' and 'temporary' entities in the grand scheme of our universe.

Any further discourse herein is best left to the philosophers! It is doubtful AI will give this, my discourse, much consideration, and EI certainly will not!

Chapter 5

Considerations Upon c (Light's Speed) And g (Gravity's Speed)

Given any dimensions within any space, there is (are) place(s). Then given time, there must be existence(s). Existence in this sense may be referenced as 'space-time,' and is then and may be described as the 'fabric' of space-time...which is otherwise not easily explained.

'Fabric' itself is an esoteric kind of term. One may espouse or conjure 'the fabric of one's imagination' perhaps? Or, a 'social fabric' perhaps? Such imaginary fabrics then become more esoteric figures of speech.

Figures of speech are, by their natures, rather vague. So, to create a more meaningful analogy to 'space-time' herein, proposed is that space-time is more like an ever-present and always-present and omnipresent 'atmosphere.' And one can imagine an analogy with ours with which to begin. Even if 'atmospheres' become cumbersome due to their specific contents, one can imagine their total presence(s), nonetheless and, as such, create the following analogies.

Here's an attempt:

1. Expansion of space-time (as ours is apparently expanding) requires some re-examination and

reconsideration…and may include one obvious contradiction for the accepted limit upon light's observed speed. Any and all objects or substances or energy waves or almost anything(s), so including even information, are understood to have a maximum speed limit of c, essentially 186,282 miles per second, when traveling <u>through</u> space-time and while <u>in</u> a vacuum. This assumes space-time to be static or essentially fixed, which it is apparently not, due to the expansion of our universe. It, our universe, and therefore our space-time <u>is</u> expanding, so then it is increasing, getting larger in 'all directions' simultaneously. It is this 'all-direction expansion' consideration that requires further definition and discussion.

2. All three spatial dimensions—height, width, and depth—actually express themselves (so are measurable) in six different directions, namely, (1) up, (2) down, (3) left, (4) right, (5) forward, and (6) backward. If our universe is expanding at rate X, from any point, say location Y, and is doing so again in 'all directions,' then X Y must be subject to all six directions…correct? However, perhaps <u>not</u>. Perhaps <u>in</u>correct(ly)…as follows.

3. Given space-time expansions only, as it is in a static state and is <u>not</u> contracting, it is difficult to imagine and impossible to explain direction number 6 = backward. Shine a light, so send a photon up off the surface of the Earth from any side or from any point on Earth or in space, then directions one through five would apply, again like that of a remote light-bulb or any star. Direction number six, so backward(s), would not be involved. In fact, direction six would

not be involved unless space-time were theoretically to be contracting…not likely, except possibly just prior to or at a Big Crunch! TBD.

4. Space-time expansion, therefore in <u>almost</u> all directions, appears to include five, not six, of them. The measurement within depth, so between forward or backward, depends upon any state of universal expansion greater than a contraction at (or possibly in excess of) lightspeed. Again, our universe, so our space-time, is expanding, and direction six must be 'out,' as in not involved. "All directions" seems to include five, not six, in our universe's current space-time existence, therefore.

5. Assume for the moment that Earth's atmosphere is analogous to a static 'terrestrial' space-time, atmosphere, and that it can be manipulated or moved in any direction, so in all six—for the specific benefit(s) of the four examples below. Examples may be many, therefore, but consider just these four herein:
   a. baseballs
   b. bullets
   c. airplanes
   d. pelotons

6. For the moment, exclude friction, specifically Earth's existing 'atmospheric friction,' which would apply. Baseballs, bullets, planes, and bicycles all deal with pushing aside atmospheric/gaseous molecules for sure. Some evidence may be found that 'frictional' (not fictional) considerations may apply in deep space too. But for the moment, and to maintain the analogy (or analogies), all effects of our known

atmospheric or other friction are being purposely ignored.

7. Baseballs: 'Sly' Young pitches a fastball at exactly 100 mph. Given a stadium with still air (most are), so then with static space-time (the analogy) and ignoring friction, Sly's fastball will pass through 'Slugger's' strike zone (or not) at 100 mph, again relative to a static, unmoving atmosphere = space-time, and relative to the ground and then also to Slugger's perspective. Move back to (previously windy) Candlestick Park south of San Francisco. Give Sly a 20-mph tailwind (most exactly coming indirectly from center-field) which would be the equivalent of space-time expansion in the direction of Sly's pitching, and his fastball will pass Slugger in excess of 100 mph, relative to the ground and to Slugger's perspective again. Sly's tailwind (same as spatial expansion) toward Slugger will require Slugger to begin his swing a fraction of a second sooner to be successful.

8. Bullets: Consider any form of long-range shooting. A military sniper's viewpoint, more specifically his aim, would be most relevant. In still air (so static space-time) and given a flat (so ± level shot) and a fixed (unmoving) target, the sniper's up-and-down considerations are due exclusively to his rifle's ballistics (most rise initially), the range (so distance), and gravity...not so much to any analogous or lateral space-time effects. Nor would tailwinds nor headwinds, so in-line changes in space-time, have much effect at all. This would be the case, given muzzle velocities approaching 4,000 feet per second, ($\pm.7476 \times 3600$ seconds) or 2,727.27

miles per hour…and again, no friction. Space-time considerations herein most relevant to the sniper's aim exist in the left and right aspects…so in the "windage" considerations that apply. In this case, the atmosphere's lateral movements, so then relative to space-time 'crosswinds' aspects, at (a) the muzzle, (b) the mid-flight, and (c) the target do apply. Modern rifle sighting allows for adjustments within the rifle scope's optics, wherein the 'crosshairs' (properly the reticle) may be adjusted up or down and right or left. So the rifle, when fired, will not point exactly to the target. Similar adjustments with moving telescopes must be made by astronomers to include "proper motion" when viewing distant, cosmic light sources over time. So distance subject to gravity (elevation or drop) and aspect (but we have specified a flat, not uphill nor downhill shot) and windage (so lateral 'space-time-like') considerations must apply to/for the bullet's path. And all apply to being on target, and a bit like Sly's aim within the strike zone, if the winds were to shift laterally either in Candlestick and downrange. Any lateral space-time movement (the analogy again) must affect any object's or particle's (as in any bullet's) direction(s).

9. Airplanes: Some specific constant(s) must apply. As an example, given a Boeing 757B that cruises in still air at exactly (so with standard throttle) 550 mph, then its air speed and ground speed would equate. Disregarding this 757's takeoff and landing, so then at elevation and at cruising speed of 550 mph, a destination some 1,100 miles distant will require two hours' flight time. Enter "winds aloft," being the same as space-time considerations in our analogy.

A headwind at 100 mph (not uncommon) reduces the aircraft's <u>ground</u> speed to 440 mph and increases flight time from 2 to 2½ hours…again, with a fixed throttle. Conversely, a 100-mph tailwind would decrease flight time to 1⅔ hours. Winds at angles/ sidewinds complicate the 757's directional settings, but the analogy fits insofar as winds aloft, so analogous to all six possible directions of nonstatic space-time must apply. A tailwind would be comparable to our space-time expansion, therefore, and vice versa. With a 100-mph tailwind, a Boeing 757 would be flying at ±650 mph relative to an observer on the ground.

10. Pelotons: As in groups of cyclists…think of the *groupo* in a Tour de France race stage. Those riders "up front" are pushing the atmosphere, so accelerating space-time in our analogy, and create a "wind shadow," so an actual moving partial vacuum behind themselves. Enough lead riders will move enough atmosphere (thus affecting a forward velocity of space-time) that riders 'in the pack' or 'off the front' but still in the *groupo*/peloton behind can maintain the same velocity with considerably less effort(s). Moving in or with the pack in a competitive situation such as in a Tour de France peloton is said to reduce a cyclist's required workload by as much as 30 percent! The forward-moving vacuum in a sense mimics expanding space-time. A contracting space-time would evidence itself as headwinds at the front and slow the entire race accordingly.

11. So what might the above atmospheric examples mean to a photon actually in a vacuum state and in our <u>actual</u> and <u>expanding</u> space-time? A photon's

direction is always forward relative to its source or origin…unless reflected, <u>but</u> <u>also</u> is subject to and affected by gravitational lensing (think windage, range and gravitational effects for a bullet or sidewinds aloft for an aircraft), and then always to direction 5, movement forward, and so then to its constant velocity. Again, as our universe, therefore our actual space-time, <u>is</u> expanding, then any backward aspects, so direction number 6, "backward" would seem not to apply.

12. So now consider an observer outside of our universe. (No concerns exactly where he or she might actually be!) Or consider an observation point fixed within the universe but not subject to any universal expansion (however difficult to assume). And assume also for the moment, at a specific distance from Earth ('where' the relatively fixed observations can be made), that exactly there, universal expansion is occurring at one-third lightspeed…which is not much different from some current observations. Regardless, the off-Earth, especially the out-of-universe, observer(s) would observe our Earth-sourced photon(s) at 300,000 kps normal velocity <u>plus</u> the 100,000 kps (so one-third lightspeed) universal/space-time expansion, or at a combined velocity of 400,000 kps if/when in the same direction. Note only, <u>any</u> velocity over 300,000 kps (so certainly 400,000 kps) is "not permitted." And this would be a contradiction. Therefore…

13. Lightspeed, c, seems correctly to be a limit <u>through</u> or <u>across</u> space-time only, so excluding the expansion (or possible contraction) thereof. Lightspeed within space-time must give c a relative "boost"

in the same direction as its expansion...or slow it similarly if/when contracting...either being relative to observation not affected by said expansion or contraction.

14. The analogies above, say baseballs to bullets to Boeings to bikes, may be applicable or not in all senses. A photon would make no 'effort' to exceed its own velocity with, so in the same direction of space-time expansion, but would logically be accelerated by said expansion as seen by an unaffected outside observer. Were space-time ever to equate or even exceed c in either direction, photons traveling in the opposite direction would therefore be observed to stop or even to reverse themselves...potentially? All this may be a fanciful "mind experiment" until someone can figure how to observe either outside of, or to be unaffected by, space-time. And then, to think about any travel at or in excess of c is troublesome at least. But...

15. According to the mathematics of distant "quasars" that seem to be receding from Earth at "near lightspeed," they may, given the addition of space-time expansion, actually <u>be</u> exceeding lightspeed? Does this continue the possible contradiction? Hubbell's law, that the speed of our universe's expansion increases mathematically with time (so distance), implies that said excess will be unavoidable...again, at some future time. When (if ever) can spatial expansion cause relative velocity to exceed net c for any object therein?

16. And what about gravity? What about gravity's theorized gravitons? They, like light's photons, must travel at c. But gravity would be (is?) seemingly

unaffected by space-time expansion. Such deviations in gravity's path (so lensing) would not apply. Gravity is gravity and logically would not lens nor be lensed itself! Photons represent a radiant force that is generated from a point, so photons do travel initially outward from their source. Gravity seems to operate in the reverse, always in a sense flowing inward toward the center(s) of any relevant mass(es). Any two reference points with mass would most likely confirm gravitons traveling 'in reverse,' therefore toward themselves. This may be number 1 (of many?) fundamental difference(s) between photons and gravitons...being that their relative directions of travel are opposite. TBD.

17. Light is also slowed by any transparent media through which it passes, therefore other than in/through a vacuum. Again, when in a vacuum lightspeed = $c/1$. In Earth's atmosphere = $c/1.0003$. In clear glass $\approx$ $c/1.5$. In water $\approx c/1.75$ (depending upon depth)... and so on. Gravity seems to flow at $c/1$ anywhere, therefore in/through anything and at any time(s)... whether in a vacuum or not. Is it simple to say then that: "Gravity speed, g (not lightspeed c), is the better universal speed constant. However:

   a. If/when and however c were to be exceeded, the relative expression of time would seemingly have to flow in reverse (an impossible contradiction, nonetheless).

   b. If/when and however c were to be achieved, mass becomes infinite, a similar impossibility. So then...

c. If/when and however c were to be exceeded, then gravity would logically not flow in reverse, but instead increase, surely.

d. Is the concept of <u>infinite</u> <u>mass</u> essentially the same as <u>infinite</u> <u>gravity</u>? (Refer to chapter 6, "Almost Infinite Mass at Almost Lightspeed.")

e. Then does gravity, like mass, compound at c? Is there any possibility of (or any sense in) a concept of mass or gravity at c+?

f. Is there any concept of <u>I</u>nfinity (so "I") becoming I+? At c+? (Refer to chapter 17, "Dimensional Proportionality.") Our mathematics to date does not seem able to explain this possibility.

g. Amounts/measurements and distances either in excess of infinity on the upside, or likewise less than infinity on the downside, would seem plausible and necessary. Are our current mathematical concepts possibly inadequate?

h. Enter fundamental difference number 2:

The "rolling ball on the trampoline" image/analogy demonstrates graphically gravity's effect upon space-time. (Or space-time's effect on any mass?) Either way, it is a rather simple to draw and simple to understand <u>two</u>-dimensional analogy. Given instead all <u>three</u> spatial dimensions, so adding depth, the diagram becomes enormously more complex. The two-dimensional plane, better, <u>brane</u> of the trampoline, becomes present in/at all aspects and at all angles, so then in a b'zillion (easier to contemplate) branes all at once. The trampoline could have to be drawn in a b'zillion (again) branes all at once, therefore…not easy to do on a two-dimensional surface! Perhaps *Star Trek*'s navigational holograms would be much easier to visualize? Prediction: Projections of such visually 'correct' holograms are not far off to affect necessary three-dimensional imagery.

18. Interesting comparison(s) exist here between light/ lightspeed, however affected, as opposed to gravity/ gravity speed, however unaffected. And it will surely depend upon...

    a. the isolation, measurement(s) and precise description of a photon, this as compared to...

    b. the same isolation, measurement, and description of a graviton, if and when observed.

       Photons, so leptons, seem to interfere with and must therefore be affected by other leptons, therefore with and by electrons (and presumably positrons) for example. It's a safe bet that gravitons, when actually described, neither interfere with nor affect any other leptons, so presumably they do not interact with any subatomic structures at all...again, fundamental difference number 2.

19. So how is it and why can lightspeed and gravity speed be so arbitrarily equal? Lightspeed can change to become both slower through transparent media or perhaps faster in line with expanding space-time. But gravity...not so much, or better not at all. Nothing, no media nor any phenomena currently understood, either slows down or speeds up gravity. Again, gravity speed would seemingly be a more dependable universal constant than is lightspeed. Difference number 3 becomes apparent.

20. "Antigravity"? A machine or propulsion system of such remains science fiction. A "counter-force" to gravity would seem to be an oxymoron. Reversing or possibly reflecting gravity, something that can be done easily to/with light, is conceptually impossible with gravity. Like the bowling ball analogy, wherein gravitational mass is being somehow affected or absorbed or refracted to be redirected with no outside influence...not likely. So difference number 4.

21. The two most obvious (and nonnuclear) forces, light (better, electromagnetic) and gravity (so then gravitational) seem to share the same speed(s) of c in a static vacuum. But, that is where their similarity ends, perhaps. When space and, therefore, when space-time is expanding as is ours, these two forces seem to have to behave quite differently therein.

Chapter 6

An (Almost) Infinite Mass at (Almost) Lightspeed

The hills behind Stanford University provided a great space to ride motorcycles. This was the case back in the late 1950s and early '60s when we were there in school as undergraduates. There was little or no traffic with which to 'compete' in the hills, and lots of dirt roads and rideable trails existed conveniently close-in to campus.

We could cruise up to the radio-astronomy disc in ten or twelve minutes, maybe a bit less given 'prime' conditions (no campus cops), from the Beta house, our fraternity, on Campus Drive. Old history now, since most fraternities are gone, but 'our hills' then were conveniently in 'our backyard.'

We all had our various access points as well...and kept them, as such, to ourselves. There was a sense of 'permitted trespass,' though no initial restrictions per se existed. The only encounters of note were with the occasional 'horseback' set that would appear almost mysteriously upon English saddles and from some road head or possibly stables of unknown location(s). Horses did not mix well with big, loud motorbikes, as ours was. So proper trail etiquette (ours) was required.

Tom Weiser and I had purchased our 'beast,' a used 750cc BMW that was overpowered (we were novices) and cumbersome compared to today's machines. But our BMW was (it felt like) pure 'macho' nonetheless...a pure 'power-trip' to ride, and at <u>un</u>safe speeds, all over our 'hills.' Frankly, it remains a small miracle that neither of us was severely injured while on that motorcycle.

Its demise was my fault while descending from the top of Alpine Road. The switchback corner I missed was severely banked, and my entry speed was much too fast. My inside, right footrest hit the road, the bike pivoted, and I was ejected...into a patch of poison oak. My unprotected arms and one side of my face sustained minor damage (road rash), and the oak's caustic effects were evident and uncomfortable for weeks. Other than that, no joints were overextended and no bones were broken. In retrospect, I was lucky.

The bike, however, did not fare as well. As it slid down Alpine Road, absent me and rotating upon its right side, the fuel line, also on the right side, unhitched and...and the bike caught fire and burned up. Nothing I could do! I had to, after assessing my own situation, regain the road and watch it self-immolate. Fortunately, no other property or real estate caught fire.

Frank Cameron, previously unknown but then and thereafter a close friend, picked me up and let me shower and change. He, at six-feet plus, had some (some kid's?) jeans that fit (I was barely five-eight) and a t-shirt that did not. My clothes, well-coated with poison oak and gravel, went in the trash. I brought him back a "thank-you" bottle of 1945 Volnay (French burgundy) from Esquin Imports in San Francisco. Frank, his wife, Suz, and I shared it with rib steaks. Delicious it was and they were.

The country road maintenance guys found the charred remains of the beast, plate intact, tracked me down, and billed me $235, big money then, for its removal and disposal… which I paid…grudgingly. I'm not sure what my attendant mindset was…that it might just be found and buried by someone else or found and refurbished by some fictional good Samaritan? It would likely have been a $235 cost or more, regardless! It really hurt, both then and now…financially and philosophically. Our 'beast' was gone, nonetheless.

The first actual restrictions were posted in the hills in late '61, as I recall, "No Motorized Vehicles." Our bike, then in existence, was unquestionably a "motorized vehicle," and this had been twice before confirmed by the same campus cops. They had intercepted us both, Tom and I, once each, about our speed and noise and illegal parking et al. So…

Tom, previously a US cycling team (track) member, opted for his road-bike and pavement. I put some fatter, low-pressure tires on a junker town-bike (pedal powered) and, unknowingly perhaps, created a primitive mountain bike. So we both went "nonmotorized," each for our purposes, as the new signs in our hills had ordered.

Next, in early '62, came more specific signage: "No Unauthorized Access." I assumed I/we all, as bona fide students, had implied "authorization." After all, we'd paid (actually, our parents had paid) our substantial tuitions, which we'd assumed included our access to our hills (on Stanford's property), right? Wrong!

"Restricted Access" and "No Trespassing" signs and fences and gates with locks all around presented new barriers. These were yet avoided with a bit of creative route finding, since the previous and obvious entry points had to be abandoned. But a 'new way' in was discovered, through private property, 'sorta.' Due to ample hoofprints and horse poop, we'd discovered the previously mysterious equestrian access! It was a kind of 'last resort' mystery trail entry that was most satisfying. The 'hills' were, and needed to remain, our terrain. But…

I was kicked off for 'trespassing' shortly thereafter by 'authorities' of then questionable 'authority.' Imagine that… trespassing while still 'on campus,' however extended. I was still on Stanford's property, but not so permitted. Seemed unreasonable. The trails and terrain were all there, still perfect as they had been for our nonvehicular and nonimpactive purposes…but forbidden still. Something "was up." Total, unexplained denial of access made no sense. Until…

"Construction Zone—Hard Hats Only Beyond This Point." This was pretty clear…except that there were plenty of guys, surveying and setting stakes and flags for whatever, also behind the signs and without hard hats. Rich Garlinghouse had a beat-up yellow one, which I borrowed… and it allowed me to periodically check out the unexplained construction, either through or over the gates (on weekends), but always on foot.

Anyway, I was able to poke around in my yellow hat. About then, I also had to cover Tom's one-half cost of our bike, which at that time was evidenced only by a charred, black spot on Alpine Road. But a bit above that spot, I could look down at the layout and the first dirt movement of whatever was under construction in 'our' hills.

I saw that a long, narrow slot had been rudely gouged out. It appeared to be over a mile long and pointed north, roughly north toward San Francisco. And upon closer attention onsite, the surveyors were using lasers on complicated tripods instead of more traditional 'bubble' leveling devices. The site might well have been a launching (or landing) strip for some kind of aircraft or rocket of some sort, except that it was far too close to the main Quad/campus.

So I simply asked one guy who appeared to be in charge: "What's up? Why the lasers?"

He was an informative sort to a point and explained that the extra-long 'pad' had to be exactly flat so to avoid even the curvature of the Earth. He went on to say that it did extend, start to finish, roughly one mile, and that it would provide the base/location for a "sort of ultra-high-speed experimental device." Any more than that, he suggested I contact someone at Stanford's department of particle physics…which I did. The 'mystery' project was far too delicious, and my curiosity was now well-established.

Under construction was Stanford's very own and state-of-the-art linear particle accelerator…acronym, SLAC. I learned that other smaller versions existed already at Lawrence Livermore Lab at the UC Berkeley campus and at Los Alamos National Lab in New Mexico. I inquired and found that still others were operating in Europe and in the Soviet Union as well. But this new one, right here behind Stanford, extending a mile, almost to the intersection of Sandhill and Woodside Road(s), was "state of the art"… and that it was a joint (jointly funded) project between the university and the US Atomic Energy Department. So, heavy company and some secrecy were apparently involved.

Now the "No Trespassing" signs made real sense. A world-class "atom smasher" was under construction. Our bike trails had been given a legitimately higher purpose.

"Atom smashing" was then, early 1960s, not exactly common lexicon, even in our undergraduate science classes. But the mechanics thereof were extraordinarily interesting… actually, they were conceptually exotic to me at the time. Stated simply: "Heavy particles," notably a stream primarily of protons, would be accelerated to near lightspeed and targeted to "split" the nucleus, apparently "one only (?)" of various heavier elements. These targets again, being the exact centers of such heavier elements' atomic nuclei, explained the laser-based accuracy requirement of construction.

Targeting of any atomic nucleus with protons or neutrons did require accuracy that would be affected by the Earth's curvature…if even over (just) one mile. This made sense. But the details of a machine able to do this remained exotic and relatively unexplained.

The machine was to be operated right there in the hills behind Stanford…remarkably. So, during the process of its construction, I tried to understand more exactly how it would work. The core was described to me as a "tube, about four inches in diameter." The "tube" ran inside a "pipe of solid copper with walls several inches thick." I never confirmed this exactly, but it made sense. Again, the tube had to be laid-out and aligned laser-straight and so aligned perfectly for roughly a mile…quite some engineering. And its atmosphere within would have to be removed, so that the 'empty' space that remained inside would recreate the "conditions of outer space"…again a perfect (as possible) vacuum, and again extant in Stanford's backyard. Scientifically "sexy," this was!

The removal of atmosphere required vacuum pumps to be located "roughly every two meters" and effectively for the entire length of the machine. I figured about eight hundred pumps would be installed. I was informed that any occasional and obviously unintended collisions with errant, so nontargeted atoms, typically of nitrogen or oxygen that might remain in the tube during operation, were problematic and would require temporary "shut down" for the machine to "cool off." So efficiency of the vacuum pumps was "critical."

Also, the acceleration of the particle beam, of protons or neutrons, required "a lot of energy," which was to be supplied by "cyclotrons" to be located in sequence apparently in-between the vacuum pumps, also for the entire length of the machine. So, add another eight hundred. Said cyclotrons would create magnetic fields that would accelerate the protons to within "99.9 percent of the speed of light." It was not explained how neutrons, with no electrical charge, might be so accelerated. Regardless, the protons would arrive with incredible mass (known to increase to "infinite mass at lightspeed") at the target(s) right there, some three hundred or four hundred yards from the Woodside Road intersection. Some local residents thereby got 'wind' of this, and they became nervous. Also, during construction, design improvements were made to increase this speed 'tenfold,' so to 99.99 percent of lightspeed. Hard to imagine. Now the neighbors were understandably <u>very</u> nervous!

The intended collisions, any one, would "smash" or essentially "explode" the targeted nucleus and therefore release most all the <u>sub</u>atomic particles within. Most then were still theorized to be necessary to keep said nuclei organized and to keep the nuclear contents, known to be collections of stable protons and neutrons, structurally intact. In this explosive process, said subatomic particles, most that had not yet been confirmed by observation, might actually be "seen" and so <u>actually</u> observed…this being the intended purpose of Stanford's machine.

It was unique to begin to understand the operation and modern technology of this advanced machine, right there in the hills where we used to operate our comparatively primitive motorbikes. Again, the hills <u>had</u> been assigned a much greater purpose!

Observations at the target end toward San Francisco would be made then in a "vapor chamber" in which the collisions would occur. Current, newer machines like the large hadron collider at CERN have replaced said "chambers" with advanced, real-time digital-read-out devices that can measure trajectories, masses, velocities, and lifespans of subatomic particles, so released, almost simultaneously. As opposed to…

Back at Stanford's SLAC, all data had first to be essentially physically (photographically) measured and then manually calculated. Said new accelerators, like CERN's are, now sixty-plus years later, so much more sophisticated and so much more powerful…in hindsight. Hindsight was not a factor, however, back then. It was "cutting-edge" science at the time in the 1960s in the Stanford hills.

Stanford's vapor chamber, and this is per the particle physics department's explanation, was to be filled with a noble gas, argon, I recall. Argon would essentially "phosphoresce," so produce light tracks with the passage of the subatomic particles so released. High-speed, time-lapse photography would record said passages as linear for uncharged particles and curvilinear images for charged particles…the chamber being subject to a powerful magnetic field. The images of photons released by the passages of subatomic particles, not of the particles themselves, were to be recorded and measured.

Some of these early photographs professed to have shown the passages, therefore the evidence of "gluons," I recall. But again, since some of these subatomic structures approached the theoretical dimensions (then) of photons anyway, the evidence of their existence(s) would be observed again as passages, so then of "passings" only…most of which existed for milliseconds only. Subatomic particles, then theorized only, were 'busted out' of their atomic architecture and exposed and so photographed almost instantaneously. It was "great stuff" nonetheless. Cutting-edge science again in our hills.

Knowing that both protons and neutrons do have similar masses, arbitrarily assigned as "atomic weight(s)" of <u>one</u> (note only: neutrons have been shown to be a bit heavier since) by Dr. Einstein's and more currently Dr. Murray Gell-Mann's and others' calculations, either would have to be approaching (almost) infinite mass when at (almost 99.99 percent) lightspeed. What an awesome event, however described and however managed! It was again of prime concern(s) of those living at or around the Woodside Road intersection. They objected to the possibility of themselves and their properties being "consumed by a nuclear explosion" that was surely "to be created" by the collisions within the vapor chamber, etc. Remarkably, nothing came to pass to change the machine's location nor its endpoint. Said folks, so concerned and so objecting, asked for and apparently received the assurances of the AEC (Atomic Energy Commission). The actual 'satisfaction' of such assurance was, unlike the subatomic passages nearby, entirely <u>un</u>measurable.

But the concept of "infinite mass," however described, and frankly at <u>whatever</u> speeds might be achieved, was troublesome then and remains so yet today. The mathematics that "prove" infinite mass remains <u>entirely</u> <u>conceptual</u>. A simple "two plus two" can be shown to "equal four," which is easily demonstrated with two sets of two tennis balls or two apples or two sets of 'anythings' upon any flat surface. But an object of any known substance or construction or internal architecture cannot be demonstrated to become so massive, therefore <u>actually</u> and <u>not conceptually</u>. Again, for a proton to actually become 'infinitely' massive (so 'heavy') while at the same time to maintain its same size, however large or small, was then and remains theoretical. And…

Protons, specifically, are very small, measuring some 10^{-13} centimeters. So then, how can one or however many proton(s) possibly, <u>actually,</u> again, not mathematically and <u>not</u> <u>conceptually,</u> achieve <u>almost</u> (at 99.99 percent of c) or <u>actual</u> infinite mass (at 100 percent c)? The mathematics may be explainable, but the logic is not.

The argument herein is as follows:

1. If the/SLAC's tube was indeed <u>empty,</u> as within it was a perfect vacuum with nothing inside, and
2. And one proton is itself, at 10^{-13} cm consists of 'matter' and of nothing else, and...
3. If it can be or is accelerate(d) to c, then...
4. Nothing will happen to change the shape or size of the proton. It cannot simply/just 'change' its mass (weight) by itself. So then...
5. <u>Something</u> else must be present in the tube. <u>Nothing</u> can and must be and must result in <u>nothing,</u> including any increased mass, this regardless of velocity. So...
6. Increased mass cannot be achieved 100 percent 'solo.' Therefore...
7. To 'actualize' or better to 'satisfy' the math, a process, an interaction with <u>something,</u> must occur.

An object, so a proton, that would have to increase its size proportionally larger to effect or to represent an infinitely increasing mass obviously would not and could not fit into a tube some ±4 inches in diameter. So to affect or actually to achieve infinite mass <u>requires instead and only an increase in density</u>. Density increase must logically result from either:

- <u>Compression</u>...so then the physical reduction in size of the proton, or

- <u>Accumulation</u>…so then the addition of substance or compounding or increased concentration of internal structure(s) of the proton itself.

Logically, for <u>any</u> structure itself, be it a particle or a universe, to increase in mass without a proportional increase in size requires an increase in density. Regardless of the (conceptual) mathematics, the actual, physical change, so the increase in mass of a particle in an accelerator that approaches lightspeed must be explained somehow by increased <u>density</u>.

It remains an intellectual necessity to try to understand the actual, physical processes that occur to verify the mathematical explanation or 'proofs' that they do. One very basic process to reconsider would be the Big Bang, so then the allegedly hugely <u>compressed</u> creation of our universe. That it, or that some compressed state very much like it occurred is obvious. Even 'when' has been reasonably well computed mathematically at ±13.7 to 13.8 billion years ago. Unfortunately, no one is likely to be able to recreate the conditions, nor to be able to go back to visit, nor to observe exactly <u>what</u> happened, nor <u>where</u> nor <u>why</u>. Steven Weinberg's elaborate calculations, together with today's cosmic microwave background temperature, currently all imply and perhaps confirm this compression-to-expansion event mathematically. But its actual <u>proof</u> will likely remain elusive for all time.

Lawrence Krauss's *A Universe from Nothing* and many other works by renowned author/physicists the likes of Brian Greene, Steven Hawking, and certainly Dr. Einstein himself attempt to explain the process by which extremely compressed, energetic forces alone (so then '<u>nothing</u>' of substance or mass) reorganize and unite to become <u>matter</u>

(so 'somethings' with substance and mass), all effectively with mathematical proofs. It must be correct to <u>know</u> <u>as</u> <u>fact</u> the physical processes that actually occurred during our universe's "inflation" and "recombination," for example. But again, these actual processes remain and may always be theoretical and conceptual concerning the origin of our universe.

One recent and comparable example may have been the theoretical (computer-generated) evidence of the Higgs boson. A <u>boson,</u> a force particle typically without mass itself, suddenly exhibits itself as a particle, nonetheless <u>responsible</u> for mass…seemingly a contradiction?

Such was predicted and essentially confirmed originally at CERN. But again, bosons (messenger, massless particles) <u>with</u> mass are seemingly as contradictory as would-be fermions (massive 'matter' particles) <u>without</u> mass. Some confusion seems to remain that will require a <u>more</u> energetic particle accelerator. TBD.

CERN's device is circular (not linear), roughly fifteen miles in circumference. So <u>at</u> lightspeed c, a photon would orbit CERN's machine roughly $186,282/15 = 12418.8$ times <u>per</u> <u>second.</u> At ±99.999 percent of c (so roughly CERN's capability), the roundtrips might reduce to 12418.7 times… so slow by ±1.5 miles or 7,920 feet. But, to speed up a proton to 100 percent of c requires an infinite amount of energy. Not available, obviously.

CERN's machine may be able to accelerate particles in two opposite directions, clockwise and counterclockwise, and then collide them. But that only doubles its capability and still is not enough to equate c. So the possible examination of subatomic partners, possible 'super partners,' possibly quarks or even 'strings,' when otherwise/mathematically confirmed, remains some time off, nonetheless…

A second consideration would be to confirm exactly what happens or what is going on at/in the gravitational center of a black hole, so then at/in its "singularity." When the associated math goes infinite or is impossible to explain or impossible to confirm, then (more) contradictions arise. Attempt to divide any amount or number by zero with a TI36x Pro calculator, and the result is "error." IBM's "Big Blue" would produce the same answer, likely: "not permitted." When attempting to measure a "dimensionless point" as String Theory does (Refer to chapter 14, "String Theory."), the simple answers and measurements remain theoretical at best. Current mathematics essentially "fail" at certain levels. Singularities exist but are "not permitted," therefore? A <u>contradiction</u>.

Another consideration might be Dr. Einstein's famous equation, $E = mc^2$. (Refer to chapter 10, "The Tennis Ball Analogy.") Given a proton with mass (assigned as) one and the speed of light, known at $\pm 300,000,000$ km per second, the math is relatively simple: $1 \times 300,000,000^2 = 9 \times 10^{16}$ atomic units per kilometer per second squared. Okay, then exactly what has been 'proven'? What exactly is 'known'? A particle/proton in Stanford's machine in excess of 99.99% of lightspeed might (does?) have a mass of what, $10^{200} \times 99.99 \times (9 \times 10^{16})$, a nonsensical number at best? It's hard to know exactly what is meant by a "meter squared." So 39.37 inches squared equals 1,550 square inches, or a kilometer squared equals whatever. But what then exactly is a second (or an hour or a year) squared? Time 'times' time = ? Mathematics not related to a very specific measurement or specific occurrence subject to some <u>form </u>of measurement(s) remains essentially conceptual and essentially theoretical by its nature. Time times time is not easily explained.

One can, however, perform a relatively simple mind experiment involving one proton (in a proton stream), but alone and by itself (absent any attendant electron), then hurtling down the SLAC machine's tube at near lightspeed. What is, in fact, going on? What is, in fact, 'happening' for it to be actually approaching infinite mass? Again, its size cannot and has not increased, so one other cosmic example may provide a partial explanation.

When slightly larger, but still medium-sized stars (±5 to 10 times the mass of our Sun) run out of their nuclear fuel that is able to be fused, they expand first (as "red giants") and then collapse (the physics is complex), but they collapse to a specific density and then explode as supernovae. These events are extremely violent. All matter outside of such a star's core is blown asunder and, so subject to the intense shockwave, forms heavier elements that are distributed outward into the surrounding universe. However...

The core is effectively blown inward, or imploded so violently and forcibly that all protons within all atomic nuclei merge with or 'absorb' their attendant electrons to become neutrons. Other than a slight normal increase in mass, the protons, all of them, convert to neutrons, very tightly compacted and compressed. And the result is and becomes a very exotic and very dense celestial object, currently referred to as a "neutron star."

Several analogies have been made that "A thimbleful of such neutron star 'stuff,' if spilled upon the Earth's surface, would plunge to and through Earth's center in ever shorter and shorter oscillations to settle ultimately at center." Neutron star 'stuff' is incredibly dense so is incredibly massive as compared to...

Just normal or 'regular' neutrons. Most are bound to protons within atomic nuclei as elements or as isotopes. Elemental helium, for example, contains two each, two protons and two neutrons. Said neutrons exist 'happily' with assigned mass(es) (atomic weights) of one (currently 1+) each. Were one able to isolate these 'regular' neutrons and so then be able to accumulate enough of them to form another thimbleful and to spill it upon Earth's surface, it might just as well float away. It would <u>not</u> plunge to the center of the Earth and oscillate back and forth as a thimbleful neutron star 'stuff' might.

Neutrons, when formed in the cores of collapsed stars, are protons that have merged with electrons and condensed. They have been, however, compressed due to the force of the core's collapse. One may assume that their 10^{-13}cm dimensions, originally more or less the same as a proton's, must have reduced. Whatever…what <u>other</u> than one electron (of very small, so .0005446 atomic units) has been absorbed is not easily explained. One may also assume, however, that a neutron's mass of 1+ in a helium atom must be much less than the said mass of a neutron in a neutron star? If so, then further 'internal' or 'structural' examination would seem to be necessary.

Both neutrons and 'naked' (electronless) protons are again theorized to be constructed of three-each "quarks." Why "three-each" and why in specific combinations of "up-quarks" and "down-quarks" remains one of the existing laws of nature that require acceptance, not explanation, here in this chapter. More discussion of quarks will appear in *Fizzicks 201*.

Quarks exist (also in theory) as 'primary' fermions, therefore as very, very tiny bits of matter, also with fractional electric charges. Their substance and particular architecture are therefore much smaller and much more exotic than are the nuclear particles themselves. But quarks are "bits of matter," and they "take up space(s)" themselves as well. It would be logical to assume that quarks, again the contents of neutrons (and of protons) themselves, may be compressed into smaller and smaller spaces as well. This fits in well with String Theory, and therefore as well, with a more refined concept of space-time...insofar as it, space-time itself can be (and apparently is) expanding, so then it must also be able to contract. So, spaces even within elemental particles, nucleons, must have 'room' to, and be 'able' to be compressed, therefore.

Back inside the SLAC tube: There are no electrons (as would be attached to atoms) present, assuming the vacuum pumps are operating properly, near or at 100 percent efficiency. So any particles, specifically our proton therein, cannot merge with an (nonexistent) electron to convert. But, at near lightspeed, perhaps its dimension changes, so reduces? The only obvious explanation requires the assumptions that the (alleged now) vacuum through which it travels is not empty. (Refer again to chapter 15, "Absence of Anything.")

Consider a very small but omnipresent "vacuum energy"...this as well as it or something very much like it, or perhaps even unlike it, may, in fact, be also the elusive "dark energy." It, either as one or the same, must exist in any space, including SLAC's "vacant" tube. Assuming then it must have some sort of ethereal, let alone physical 'presence' therefore, it might be the actual "fabric of space," however unmeasured and/or unidentified currently. Some measurement(s) actually exist, so if measurable at all, it must be something. If only in an elusive form of energy (or negative energy) therefore, it may still form or present a barrier of sorts with which any object with mass (protons do) must contend...specifically when approaching or attempting to pass at lightspeed. Might our proton progressively "hit a, the proverbial wall," at c?

So then the nearly dimensionless substance or cumulative microstructures or even virtual 'stuff' present in all spaces will allow their disruption and so allow for the pass-through of any mass up to lightspeed—but when beyond, they do not? This barrier may be progressive, so it may begin to present itself when an 'intruder,' again <u>with</u> mass, approaches lightspeed. But…then it demands dimensional reductions for said intruder, a proton, to pass through? Again, the mechanics of <u>compression</u> herein and/or of accumulation (discussed below) must be involved…either way.

The more worldly notion of a jet aircraft, when approaching the sound barrier, that does experience momentary resistance that <u>can</u> be overcome with <u>more</u> velocity before going "supersonic," sounds vaguely familiar. However, a mass just at lightspeed cannot muster any more velocity and must submit…end of story. No verified "super-c" options exist in space-time currently.

So then does our naked proton compress (or is it actually compress<u>ed</u>) to a proverbial 'point' of unknown dimension to get through? Must its quarks 'cozy up' or somehow convert or realign somehow to proceed? That our obvious matter particle (our proton) of measurable size would, however, compress to a point of (currently) <u>un</u>measurable size would explain "infinite density" and so then "infinite mass."

If such an object can strike an atomic nucleus and cause it to 'explode,' rather than just pass through said nucleus and see that or allow that its 'shockwave' (so deep into quantum mechanics) does the work, is speculative at best and beyond the scope of this consideration, regardless.

So much for (a) <u>compression</u>.

The alternative explanation remains (b) <u>accumulation</u>.

Given all the possible conditions above, what again might exist <u>in</u> and not have been removed <u>from</u> SLAC's vacuum tube?

1. What of <u>photons</u>? They exist as rather persistent 'units' of electromagnetic energy. Even the (eight hundred or so) pumps must add some heat. Also, (eight hundred or so) cyclotrons produce a lot of magnetism. So photons are obviously <u>added</u> (not pumped out). Visible wavelengths would be eliminated (there is no 'light' per se inside the tube). But other electromagnetic wavelengths, so then photons, would logically persist.

2. Dark Matter, whatever it "proves to be," must exist also inside the tube. It currently explains ±25 percent of the <u>un</u>observable mass of the universe. Any considerations of 'pumping it out' are pointless and illogical. It must have some (however miniscule) mass, and would be "weakly reactive," and so then possibly accumulate within our intruder at c. Whether a waving hand or pitched baseball or jet aircraft or whatever, objects moving through space would be totally transparent to dark matter…except up to and including c…possibly?

3. Vacuum Energy (VE), also as above and <u>if</u> different from Dark Energy (DE) as below, VE would likely be represented by the, to date, very elusive and unobserved <u>gravitons</u>. Photons exist as quantum units or 'packets' of electromagnetic energy. So then must <u>gravitons</u> exist as quantum units or 'packets' of gravitational energy? Given their (photons' and gravitons') common velocities but apparently opposite 'directions,' some more complicated considerations may apply…other than simply their presences. And then lastly…

4. Dark Energy. Again, DE would likely be represented by <u>doubly</u> elusive and <u>absolutely</u> unobserved

(unobservable?) retrograde gravitons? Again, if it, DE, is responsible for the increasing velocity of the expansion of our universe, and so then of space-time, then all gravitons would be likely suspects. However described, DE appears to be responsible for a whopping ±70 percent of our unobserved universal mass. Further discussion is needed.

So photons, being massless and spaceless, are unlikely to accumulate or to somehow add mass inside a near lightspeed proton, and...

Would (does?) Dark Matter exist essentially uninvolved at sub-c, but somehow behave differently at c? The mechanics of accumulation within/inside a near or at lightspeed proton, so to create the 'superdensity' that would again equate to "infinite mass" is and are not beyond comprehension, much like...

Vacuum Energy, VE again, represented by gravitons would be most likely to accumulate and to increase mass. Assuming they, gravitons, do (must?) carry or 'represent' the force of gravity, one would expect their increased presence to generate and to result in increased mass. Again, perhaps like or perhaps in conjunction with, Dark Matter, vacuum energy would/could likely accumulate. (Note here: There exists at this juncture a possible relationship with the Higgs boson. Its (the Higgs') function, so for a boson to represent mass, sounds ever so familiar with the function of a graviton (so VE). Upon further study, such will be a fertile source of discussion in *Fizzicks 201*. Same again for the possibility of retrograde gravitons, because...

Dark Energy again would seem to require the existence of retrograde gravitons...which again logically could also accumulate to increase mass. Perhaps both 'loose' photons and 'loose' gravitons, assuming both to be present in the tube's vacuum, are avoided or 'pushed aside' but somehow absorbed by or otherwise accumulated within our proton near or at lightspeed? Again, the exact mechanics of why they are excluded then accumulated will require more thought and discussion.

The accumulation of both dark matter and/or gravitons within the structure of a speeding object with mass makes sense, however. So they become the "most likely suspects." Both must pass through any objects with mass at velocities less than lightspeed. Perhaps the absolute, maximum speed limit, c, in a vacuum (so with no other atmospheres of interference) is the result of saturated accumulation… therefore a maximum density? When there is no more space (within our proton again), then does 'maximum' equate to/ with "infinite" density? And then at infinite density, can there be no increased velocity? Is it that "full is full" and "fastest is fastest" for purposes of this discussion?

Again, the exact mechanics of "accumulation" near to, then exactly at, lightspeed needs further explanation. But the net effects of

- infinite density that results in
- infinite mass and determines
- maximum velocity…all make some sense.

All seem comfortable and 'includable' with particle physics, quantum dynamics, and relativity…as understood. So does function (the increased density) follow the form (the mathematics), perhaps?

Of note: Would the possibility of both proper or positive and retrograde gravitons then lead to a possibility of compound mass? Or might like particles, gravitons and retrograde gravitons, interact like electrons and positrons, for example? The excess of matter over antimatter must prevail…obviously. But is a concept of compound mass at all complemented by the existence of retrograde gravitons? Seemingly and preferably, they would not negate one another. The concept(s) is (are) confusing, but not illogical. Mass and anti-mass may or may not cancel one another much like positive and negative electrical charges do. Quarks with fractional but the same positive and negative electrical charges seem to function quite comfortably. Might, perhaps, proper gravitons and retrograde gravitons somehow represent even fractional gravitational forces and function much the same as each other when accumulated?

Contradictions (surely mathematical) are likely to exist/arise…also far beyond the scope of this writing. But logic has it that one is, so we are dealing with possible, better with potential, best with probable mechanics that, however it is explained, mathematically or actually, it must explain what is, in fact, going on to affect "infinite mass."

Quotations that come to mind that may provide or contain some relevance for future consideration(s):

1. "The purpose of time is to be sure everything does not occur all at once" (Albert Einstein)…quoted roughly. Time, as noted, goes to zero specifically for objects at lightspeed, so nothing can be going on… can it?
2. "No matter where you go, there you are" (Buckaroo Banzai)…quoted exactly.

Probably occupants in the vacuum of space (so of/in SLAC's tube) and their interaction(s) with massive objects, again, near or at lightspeed, must relate…somehow? And, as Mr. Banzai implied: "Ya gotta be somewhere!" And that would appear to be a requirement for anyone and anything.

I miss our old 'macho' BMW and the trails 'in the hills' behind Stanford nonetheless. The SLAC's machine, however since improved, is still in operation there. There have been no nuclear explosions nor 'black holes' at the Woodside Road intersection...nor at CERN...which would be a much more likely location for possible production of the same. Given 100 percent c for any mass requires an infinite amount of energy, photons do, and it is assumed gravitons do also travel at c quite comfortably. This is apparently no 'big deal,' because they are massless...right? Or does some other more exotic explanation exist?

Stay tuned, as will I...when "infinite" energy becomes available. The 'science' on display in our hills is likely to become even 'sexier' one day!

Chapter 7

$E = mc^2$ (Dr. Einstein)
$V = d/t$ (Mr. Euclid)

Dr. Einstein's most famous equation states that total energy, E, is equal to any mass, m, times the speed of light, c, squared. By itself and as written, the equation appears to be rather simple because c is a constant. Accordingly, given any mass m, total energy E is rather easily computed. And then also, since $m = E/c^2$, so is mass easily calculated as well.

The constant c is usually stated as either:

1. 299,792,458 meters per second or
2. 186,282 miles per second.

So for c^2, the actual, numerical constants are: 89,875,517,870,000,000 or roughly 8.99×10^{16} meters per second squared or 34,700,983,520 or roughly 3.47×10^{10} miles per second squared. Big numbers!

The equation may be written as: $E = mc^2$ or $m = E/c^2$ or $c^2 = E/m$, and either way, given any mass of one, say, one gram or one milligram or one kilogram (or whatever mass measurement of one is used), then $E = c^2$ as expressed in the same units as was the mass. Again, simply.

Energies are typically expressed as in either:

1. electron volts (eVs) = the smallest of current measurement, or
2. ergs (arbitrarily 600 GeVs), where G indicates billions, or
3. joules (an arbitrary number of ergs)...so best left to 'loftier' calculations.

Energies and their computations can be confusing. The electron volt, one eV, is the primary unit of measurement. Exact physics aside, an eV equals the amount of 'push' required to move an electron out of electron ring number one into ring number two. So MeVs = thousands and again GeVs = billions of eVs...as defined.

Again, with any mass of <u>one</u> of any measure, it is then $E = c^2$. And then, in the alternative...

As a constant, c may be expressed as one (so always the same), and c^2 is greatly simplified. When c = 1, then $c^2 = 1$. A $c^{2'd}$ multiplication when other than one results in time, typically in seconds <u>squared</u>. Whether a second or a minute an hour or a year squared, time 'times' time is confusing. With c^2 set at one, regardless, said confusion is at least less evident. And then...

"E = m when c = 1" is a very profound statement...and perhaps Dr. Einstein's greatest insight. So given again the constancy of c set at one, c^2 is then also equal to one, and any mass may be expressed in units of energy, specifically eVs. For example:

- One gram of mass equals 5.6×10^{32} eV. Notably, there is an immense amount of energy released when a relatively small bit of mass/matter is converted. One might consider the core of an early, original atomic weapon at ±22 pounds of (heavy) uranium 238, for example. Given 453.6 grams in a pound, times 22 equals 9979, say 10,000 or 10^4 grams. Then $5.6 \times 10^{32} \times 10^4 = 5.6 \times 10^{36}$ eV of energy release…and a great deal of destruction over two Japanese cities resulted…fortunately or unfortunately? It's hard to say. But a terrible war in the Pacific was ended.

- The mass of an electron = .511 MeV…and is a very small measurement. Figuring in reverse that an electron has been demonstrated to weigh 1/1836th of a proton…

- So then, a proton's mass has also been measured at 938 MeV. Note: 938/.511 = 1836 (electrons), and the eV mass calculation for an electron equates.

- A neutron's mass = 940 MeV and is a bit heavier than a proton's because of its internal structure contains two (heavier) down quarks. One may then compare 940 less 938 = 2 MeV as the quarkian weight differential…almost. And the electron volt (eV) measurement capability allows such exact computation(s) to be made.

So again, when c = 1, then E = m, and this is again a very profound statement. Further…

- The total energy that binds one electron to one proton to result in one hydrogen atom is roughly 13.6 eV. So 13.6 eV divided by 511 eV, the mass of an electron, equals 2.67×10^{-2} eV, considerably less total energy ratio for a hydrogen electron to communicate with its proton/nucleus and again at a proportionally enormous desistance. (Refer to chapter 10, "The

Tennis Ball Analogy.") How hydrogen atoms themselves, let alone all the other 'heavier' elements, stay organized is indeed remarkable.

So what? Are we or is anyone confused yet?

The Earth's weight, its mass, equals $\pm 3.4 \times 10^{51}$ GeV. Again, a GeV = one billion eVs. Just know that mass is related (in essence is equated) to energy by c^2 = the square of lightspeed. So hopefully, the Earth will never 'go nuclear'! 'Our' Sun, many millions of times the mass of the Earth, has done so...under a very controlled circumstance. Think of stars that burn all their available nuclear fuel (an event our Sun may experience in ±several billion [estimated] years) and then do explode as supernova(s). Energy releases of such events are stupendous, hard to express even in GeVs.

Understood directly or perfectly or not, assuming $E = m$ when $c = 1$, it becomes ever so much easier to quantify the mechanics of nuclear synthesis (like the Sun's mass relative to its output will define its lifetime), or even to 'reverse-engineer' the Big Bang therefore. What possibly were the parameters exactly at time zero?

To be sure, the total of all eV measurements and relationships have remained exactly the same since the beginning of our time, so for ±13.75 billion years, and have not and will not change. Photons created/released then (at lightspeed) have not aged at all. Time stops at lightspeed, so all photons are the same age now as they were then... relatively speaking.

It is indeed the past, present, and future (assumed... hopefully!) 'nonchangeability' concept that remains remarkable. Again, the laws of physics need be fixed and unchangeable...from any human viewpoint. Therefore, might our human behaviors be worth some closer examination and introspection accordingly? Natural laws (like the age of a photon) are fixed and unchangeable. Apply this concept to various recent Supreme Court decisions! Back on subject…

Noted at the beginning of this chapter also was Euclid's formula for velocity: $V = d/t$. Of note, c (or c^2) in Mr. Einstein's equation is also a velocity. So, given some very basic mathematics:

Given   1. $E = mc^2$
and   2. $V = d/t$
and when   3. $V = c$, therefore
then   4. $E = m$ (c distance/c time)^2
when c is then set at 1, then both:
                    5. $E = m$
and   6. $m = E$. Therefore…

Mass and energy must exist and consist of the same basic "stuff" at some level. And, given the constancy of mass and energy (insofar as neither can, by itself, be created nor destroyed), there must always <u>be</u> a consistent net amount of each one and/or of both. So, a state of "nothing" is both conceptually and mathematically impossible.

But look at equations number 2 and number 4 above. Two major contradictions might exist when either: Situation A' where d approaches or becomes 0…as in or at a "dimensionless point." If A' were to occur, then $E = m (0/t)^2$. Zero divided by any number (t in this instance) equals 0. Zero squared = zero. And then $E = 0$, which is not likely.

Total E equal to 0 (zero) makes no sense. So then, <u>some</u> d, <u>some</u> distance, so <u>some</u> dimensions, must exist…always. Again, where $c = V = d/t$, d must have some value greater than 0 for c and therefore for E or m to exist! There must be no end to measurement(s). Anything, so any distance, may be cut in half forever. Einstein's and Euclid's formulas seem to demand this to be true!

E = m regardless of c…period. So…

String Theory's attempt to define the absolute smallest of structures cannot be the end of all dimension. No matter how small 'strings' are ever actually proven to be, they must also be of some size greater than zero and so, logically, must have to be of some substance to exist therefore.

And major contradiction number 2 would exist in this instance…call it Situation B' wherein:

> Time approaches or becomes 0…as in when or if any/all actual time were to stop. If B' were to actually occur, then $E = m (d/0)^2$. Any number (d in this instance) divided by 0 produces the irrational answer of N.A.N. (not an answer). In such an instance as this, c itself would be at least infinite (so well in excess of lightspeed), as would E > infinity, which is not possible. Infinity 'squared' would be worse! Therefore, even an "instantaneous" event must require some time.

Neither infinite E nor zero E make any sense.

Total E obviously exists as a very large but presumably fixed amount that was determined at and initially set by our Big Bang. So then some t, some measurement of time, must exist also and always. The 'instantaneity' of any event must be questioned. Current measurement capabilities of any dimension or of any time may appear to be zero relative to the specific occurrence of any single event, but any such event must occur in/within some space-time regardless. So…

Our measurement capabilities must be inadequate. Neither zero distances (so zero dimensions) nor zero time are either mathematically or logically permissible.

Again, since space and time obviously <u>do</u> exist, then so must both d and t always have values in excess of zero. And, given the consistency of c, velocity, V must also exist in proportion to c as well for $E = mc^2$…always.

Here are summation points:

1. For E to equal mc^2, then
2. E must have some real value(s) greater than zero, but less than infinity.
3. Since c = light's d over light's t = V, then
4. both d and t must have values greater than zero. And
5. Dr. Einstein's equation in essence defines space-time's mandatory existence, therefore.

"Who can imagine a distance that cannot be shortened?"

"Who can imagine a time that cannot be lessened?"

So both a "dimensionless" point and an "instantaneous" event are not even logically permitted.

So again, the express logic of $E = mc^2$ becomes so much more evident. It describes the necessity, as well as existence, of space-time itself.

Perhaps Dr. Einstein has already stated the TOE (theory of everything)? Such a 'simple' equation it is; nonetheless and how subtle it is as well. Why look any further?

Notes on Chapter 7, "E = mc² (Dr. Einstein)"

1. Per chapter 9, "Mario's Champagne," it appears impossible to properly describe <u>any</u>thing at zero velocity…especially light. And yet, the most ingenious experiment known to this author attempts to do so. Allegedly, light's speed was reduced to zero when projected through a thorium (?) crystal and at an absolute zero temperature…so at (negative) -273° Celsius. Note that -273° Celsius is, by definition, also 0° Kelvin…and that 0° K indicates the absence of any/all energy(ies). So, light (an energy) through any (assumed, translucent) substance such as a thorium crystal at 0° K seems contradictory with which to begin. It is incomprehensible to think of photons simply 'stopping' within any crystal or any substance…at any temperature. And again, at or even close to 0° K, would light, when/if present, be assumed to pass unimpeded, if at all?

2. What about:
    a. $V \pm d/t$ when $t = 0$, so measured at an 'instant,' so
    b. $V = $ infinity, when any number (of d) is divided by $0(t)$? If mathematically possible, then V must exceed lightspeed—which cannot be exceeded. If again, a perfectly reflected (at 180°) photon must have (express) a 'moment' of 0d and 0t exactly at its 'turnaround.' But…
    c. For E to equal mc² at any/all times, both d and t, or Δd and Δt, however calibrated, both must have some value(s) greater than 0. So, both a 'dimensionless point' and/or an 'instantaneous event' seem both illogical and impossible given our current understanding(s), which may reflect upon our current measurement(s), which must imply that…

    d.   Our mathematics, as currently constructed, is (are) inadequate.

3.   The logical proposition is that here, in our world, so then in our universe, neither d (any measurement of distance or dimension) <u>nor</u> t (so any measurement of time) can be either "too small" or "too big." And this proposition has been consistent herein. So then it follows…

4.   If both d and t must always exist (that is, they must always have values greater than zero), then V (velocity) must always exist as well. For any object or event to have or to demonstrate no motion or zero velocity, its value of d must also be 0 (zero), which is not permitted. Accordingly…

5.   For c to be our universe's maximum speed limit, then all its elements must also have values greater than zero as well. With t at 0, then V equals or exceeds infinity, which surely exceeds c and is not permitted. So t <u>must always have</u> a value in excess of 0…both actually and mathematically.

6.   The existence then—or better, the persistence of motion—at all times is also consistent. (Recall chapter 9, "Mario's Champagne.") Motion, like velocity, requires <u>both</u> some distance and some time…always.

7.   So then neither E nor m can be 'still,' so without any motion, and so then without any velocity…relative to <u>some</u> observation or <u>some</u> viewpoint. Our world is constantly in <u>some</u> form of motion, any and all of it, at any point in time. Absolute "stillness" is not permitted.

8.   Subject to notes number 3 and number 4 above, simply restated:

    a.   Absolute stoppage of all motion would require 0 distance, so 0 dimension, and

b. Absolute stoppage of all time would result in infinite velocity, and

c. Neither are permitted.

9. Absolute maximum velocity c is obviously evident (so greater than 0) and is obviously limited (so less than infinite). It remains remarkable that this particular speed/velocity, whether measured in meters or miles per second (or whatever distance or time sequence is noted), is, has been, and certainly will (must!) remain exactly fixed and exactly the same for all time(s). "Who ordered that?" (not an original quote).

10. C's relationship to gravity, or vice versa, is also unique...exhibiting also the same speed and the same velocity always, when c is measured in a vacuum. To relate lightspeed exactly to gravity speed makes sense, except perhaps when compared to heat (possibly slower?) or x-rays or magnetism (possibly faster?). Not really because all electromagnetic forces or waves are transmitted at c by photons. And photons, units of electromagnetic energy, have been identified, observed, and quantified. But curiously, they have not yet been measured for size or dimension. Therefore, a photon's d = ? is another discussion. (Refer to note 15 below.)

11. If we can call gravity speed g, then c, so lightspeed in a vacuum equals g...wherever. Newton's theory of gravity assumed g was instantaneous, so then infinite. But, when Dr. Einstein added time and c's consistency, then g = c. So then may his most famous equation be written $E = mg^2$ with no change but some improvement in accuracy?

12. But said 'accuracy' is not yet confirmed because gravitons, the 'units' of the gravitational force (again, being perfectly comparable to photons being the

units of electromagnetic force) remain <u>un</u>observed, <u>un</u>identified, and <u>un</u>quantified. It <u>has</u> been confirmed, however, that photons must behave both as particles <u>and</u> as <u>waves</u>. Gravity <u>waves</u> have recently been observed and confirmed as well. So it would appear that gravity 'particles,' gravitons, must also exist, much again as photons do. Deductive reasoning should (and best would) apply.

13. What also, mathematically, must travel at velocity c? <u>Time</u>. Deductive reasoning again: At c, any object with mass (so anything with substance) demonstrates three very unique properties:

a. <u>Infinite mass</u>. But since photons have none (are massless), no problems nor contradictions exist. Photons travel easily/naturally at c. So then must gravitons have no, or zero, mass as well. They must also travel effortlessly at g, the same velocity as c. More deductive reasoning.

b. <u>Lack of d</u> = lack of dimension. So d goes to 0 at lightspeed, which would define also 0 zero velocity, the same contradiction noted in note number 3 above, except for one specific value of t…that being $t = 0$ as well. The implication is then that zero distance over zero time $= 0/0 = 1$ at c…a mathematical oddity perhaps? To be discussed further.

c. <u>Lack of time</u>. So t must also equal 0 at c. Zero behaves oddly. By definition, any number (assumed to be positive or negative) when divided by zero results in a sort of infinity and is "not permitted." However, at c, for both d <u>and</u> t, where t is for that event only, then d/t must equal one at c. So then d/t at c demands that 0/0 at c = one, not infinity. "Who figured that?" (an original quote). The math and the logic appear to be in contradiction?

14. Time t and lightspeed c are obviously (arguably?) inversely proportional. At c, time is said to stop, so then t would (theoretically) equal 0 at c. Conversely then, when c = 0, so when there is no movement at all, t must equal c. Therefore, in a sense and when all three—c, g, and t—are operating freely and unimpeded (as currently must be the case), then c = g = t.

15. Zero time passage, so zero time measurement for any particular event or any particle or for any mass at lightspeed is convenient for sure. The effects of d = 0 and t = 0 at c seem to demand it. Accordingly, all photons (and assuming all gravitons) so created during (a consideration of *t*, again?) the Big Bang are the same age now as they were then…some 13.7+ billion years ago! That would be age zero. Years don't matter for anything or any event when time = 0 for that specific thing or event. So neither photons nor gravitons age at all. Both are (and both will be when gravitons are observed) massless and ageless and traveling (remarkably consistently) at c…so then they are timeless too. Both may be observed exactly as created at zero time then, so at <u>any</u> time now! So for time's speed to be any different is illogical.

16. The thought then of measuring either a photon's or a graviton's dimension may make no sense. Simply, if they travel at c (or at g as above), they cannot have any mass or dimension, so why bother? So, both photons do and gravitons will become "virtual" by earthly definition, so thereby, they would exist as "virtual particles." But…

17. Deductive reasoning (again) argues that any<u>thing</u> must be either real or be virtual, the implication being that they cannot be both at the same time. So much

for "thought(s)" or for "information." So "virtuality" contradicts "reality" in a sense that will best be left for *Fizzicks 201*...coming later!

18. C, lightspeed, is in reality only a <u>maximum</u> speed when in a vacuum. So it is not then really a <u>constant</u> as properly defined because c, measured in any other translucent medium (so other than in a vacuum), is slower. Again...

    a.   C is measured times 1/1.0003 in Earth's atmosphere (assumed at sea level).

    b.   C is measured times 1/1.5 when through clear glass...so it is assumed to slow similarly through fiber optic cables or telescopic lenses, etc.

    c.   C is measured (estimated) times 1/1.5+ through water, and slower at depth, etc.

    d.   One alleged experiment (unconfirmed) claimed that c was reduced to zero when somehow "retained" in a crystal of thorium at absolute zero. Imagine an immobile photon? Not likely. Nonetheless...

19. Lightspeed limit c is consistent, so it is a constant only when traveling in—more specifically when traveling <u>through</u>—a vacuum (of space-time), which often it is not (refer to note 18 above). *C* slows down when traveling through any other medium...when it is <u>not</u> traveling in/through a vacuum.

20. Gravity speed, g, is certainly equal to lightspeed, c...fair enough? Again, it is "fair enough" only when traveling in/through a vacuum. <u>However,</u> g seems to be immune to any media-slowing effects at all. Gravity seems to proceed at *g* regardless, whether in/through a vacuum or not. So g would likely measure always times 1/1, and then g = g whether in Earth's atmosphere or in/through clear glass or optical cables...or whatever, yes? No medium appears to slow g, water in particular, for example.

Changes in buoyancy may occur via the displacement of any volume of water…and perhaps some effect to via the replacement of atmosphere(s), so measurement of <u>weights</u> may reduce, but not the speed of g. The speed of gravity, so then the velocity g, remains unaffected, therefore…and <u>always</u> at g whether in a vacuum or not.

21. It would appear that g, rather than c, should be accepted as the universal <u>constant,</u> therefore…perhaps for sure when gravitons are observed and properly described? Hopefully, Dr. Einstein does not mind, but $E = mg^2$ appears to be a more exact equation per any logical understanding. Objections are welcome…with proofs. Can gravity be slowed through <u>some</u> medium, perhaps? Hard to know for sure, but gravitons when confirmed will answer a lot of questions.

22. Again, the Higgs <u>boson</u> was recently, albeit mathematically, observed at CERN. It appears not to be certain whether it was actually <u>identified,</u> but it was apparently <u>quantified</u>…at 100–200 GeV! The only other bosons with mass(es) would be the W+, W-, and Z^o particles that transmit the "weak nuclear force," which must be another more technical discussion. But a boson (again the Higgs) that explains (transmits) mass sounds very much like a graviton. Further experiments/ collisions at CERN, other than the Higgs result, have been inconclusive. Is, perhaps, the most powerful (and expensive) machine on our planet not powerful (and therefore not expensive) enough?

23. But I'd opt again for $E = mg^2$ as the proper expression of the relationship between mass and energy…when gravitons are confirmed. Apologies to Dr. Einstein regardless. And…

24. When operating 'on their own' and when unimpeded by and in any circumstance(s) at all, then c must = g and must = t. How to calibrate t will be a requirement of *Fizzicks 201*. Please stay tuned, therefore!

Chapter 8

Lightspeed...Can Be Confusing

A meter was thought (and originally meant) to be one ten-millionth of the distance between the Earth's equator and (via any direct meridian) to either pole. This assumed the Earth was and is symmetrical...which it may not be given its spin (rotational) circumstance and irregular surface features, etc.

A meter is currently defined as the distance light travels in a vacuum, in $1/299,792,458^{th}$ of a second. Therefore, light's speed, correct to 10^9 is 299,792,458 meters per second... which has been rounded to ±300,000 kilometers per second for less than exact calculations...noted later.

Note: Now here on Earth, light does not travel in a vacuum. It travels in and through our atmosphere, so it is slower by 1/1.0003 (at sea level), so it travels roughly 299,700,000 meters per second...slower by ±92,458 meters. Such slowing measured instead through clear glass is 1/1.5, closer to net ±200,000 kilometers per second. Actual lightspeed depends upon the medium through which it travels. Water, for example, slows it considerably.

But true lightspeed in a vacuum (assumed to be also in/ through the vacuum of space) is, again, 299,792,458 meters or 186,282 miles per second...confirmed. Nothing, not even "information," can travel faster...can it? If Dr. Einstein et al. are correct, nothing can travel <u>through</u> space any faster. However, if space itself is expanding, the observation of lightspeed may be more complex. (Refer to chapter 7, "E = mc^2 (Dr. Einstein).")

Of note also is a yard, which <u>is</u> 36 inches. A meter is 39.37 inches. Both are arbitrary distances. So then:

> A yard is .9144 meters, and
> A meter is 1.0936 yards.

One thousand meters is 1,093.6 yards or a kilometer. A mile is defined as 1,760 yards. So 1,093.6/1,760 equals .62136 kilometers per mile exactly, and...

All the math works, except when using the 'generally accepted' measurements of "300,000 kilometers" or "186,000 miles" per second as <u>approximate</u> lightspeed. Note that 186/300 = .62, which is assumed to be miles per kilometer (same ratio). But the difference of .62136 (exactly) less .62000 (approximately) is .00136...so 136 hundred thousands, which would not seem to make much actual difference. Except that this difference amounts to roughly 25,000 inches or roughly 2083⅓ feet or .3946 of a mile per second at lightspeed. In a year (3600 × 24 = 8640 × 365 = 3,153,600 seconds), this distance differential is roughly 1,244,400 miles and obviously <u>does</u> make a difference.

Why might this matter? <u>If</u> travel <u>at</u> lightspeed is ever accomplished, any traveler, and let's call him Joe, would not age when at lightspeed c. Note that 'time speed' and 'lightspeed' are essentially opposite. At lightspeed, time's effect, so its relative passage (or speed) is zero. So then all clocks and biological processes, in fact, everything that is time dependent stops, which is a benefit for Joe. So then at stationary speed 0, time speed must equal c...another discussion. Stationary speed 0 may be impossible, however, to discuss further. (Refer to chapter 7, "$E = mc^2$.") Regardless...

An interstellar trip to the next closest star, Proxima Centauri, (not our Sun) will require ±4.4 light-years, one way. And assuming a half-year 'fly-by' only then (while rounding the star), say it would require a ±9.3 lightyear roundtrip. Such a trip would be well within reasonable observation here on Earth and again require ±9.3 years, plus roughly one year for acceleration both up to and then deceleration back from lightspeed, say to return in orbit around the Earth. This would assure easier recovery in case of any malfunction, and limit the G-forces to be tolerable by Joe. Logically then this acceleration and then the same deceleration, again each of one year (so back to 0 speed at landing) would be required similarly...so add two Earth years for acceleration and deceleration for this (or any) interstellar trip.

Joe's actual takeoff to touchdown trip would then theoretically require 1 + 9.3 + 1 or a total of 11.3 Earth years. During both acceleration and deceleration, so from 0 to lightspeed and back, let's assume Joe's age would slow by 35 percent. Recall at lightspeed, Joe does not age at all. So getting up to and back from his zero age, travel time at c will consume ±.65 × 2 = 1.3 years, which would approximate the total time for any such trip from Joe's viewpoint and/or experience. Again, while at lightspeed, Joe does not age.

It is notable here that a proposed trip for Joe to our nearest galaxy, so well outside our Milky Way, would require some twenty-five thousand light-years each way, plus a turnaround and the same one year each acceleration and deceleration.

Therefore, in Earth years, Joe's galaxy trip would require: 1 year (acceleration) + 25,000 (outbound) + 1 year (longer turnaround) + 25,000 (inbound) + 1 (deceleration) or 50,003 years at least...essentially beyond any reasonable observation. Joe would still age only 1.3 years, same as/for his star trip, and would return therefore 50,001.7 years into his future!

Assume for the star trip, Joe's age at takeoff on December 31, 2017, was 30. Joe will return at age 31.3, but on or about March 31, 2029, Joe will have landed back into his future by 11.3 less 1.3 or about 10 years. His wife, if age 25 at takeoff, so five years younger, would be roughly 36⅓ upon his return, or five years older...a bit of a shocker, but tolerable (hopefully).

Assume for a galaxy trip, Joe's age, again 30 at a December 31, 2017, takeoff. Again, Joe will return at age 31.3 but on or about 52020...52020? Yes, Joe would have no wife (having been gone for roughly 51,930 years!). In fact, there is no promise at all that there would be any recognizable human civilization at all to which Joe might return!

Conclusion number 1: Since 186,282 miles per second, or 670,615,200 miles per hour, is the maximum universal speed limit, period, interstellar (so relatively close-in) travel may be possible, but intergalactic travel will not. End of story. Recall our Sun, at ±93 million miles distant, would require less than twenty minutes, roundtrip, actual flight-time at lightspeed. If again it requires two years just to get up to and back from such a speed, why bother? Lightspeed travel, if ever achievable, definitely has its limitations!

Conclusion number 2: This discrepancy of time over intergalactic distances may explain our currently observed "silence." Why, if so many possible life-forms or civilizations do exist, even simultaneously, at this time, here, then when may we be aware of their existences? Their 'information' may take thousands of years at lightspeed to reach Earth. That 'other' civilization, say in the outskirts of Andromeda galaxy, may be twenty-five thousand light-years distant. Then any signal or any bit of information, however sent, even ten thousand years ago, may not reach Earth for a very long time from now, say ± year 17,022 per our calendar. No wonder there is 'silence.' Light (so information) is simply too darned slow!

Interesting also to note the possible 1,244,400-mile-per-year error caused by 'rounding' lightspeed to 300,000 (only) meters or 186,000 (only) miles per second. Given just Joe's star trip, such an error would amount to over 13 million miles…a possible over- (or under- if in miles) shoot. Given a (possible) galaxy trip…forget it. Joe or any future traveler would miss our solar system entirely upon return! So, space-flight engineers, take note: "Exact accuracy is required if you expect to see Joe again!"

Nonetheless, what does, in fact, travel at 186,282 miles per second in a vacuum?

1. Gamma rays—very high-energy radiation.
2. X-rays—at the dentist, in the hospital, etc. (actually, again a bit less in Earth's atmosphere).
3. Blue light.
4. Green light.
5. Red light…3, 4, 5—visible light…(ditto).
6. Microwaves (all a bit slower here on Earth).
7. Radio waves (a bit slower here on Earth).
8. Heat—yep (to discuss the CMB).
9. Gravity—waves confirmed, gravitons not (yet). But gravity must not be possibly slowed like light can be slowed in a translucent medium…to discuss. Gravity travels at c in/through all media.
10. Time. Think of it, at lightspeed, effectual time = 0. So at rest, time = 186,282 mps. Perhaps this is our maximum 'speed of thought'? If so, a major contradiction exists! To discuss also…another

'time.' How to properly 'calibrate' time remains elusive.

But, all the above proceed at the same speed when traveling unimpeded and in a vacuum…assuming a vacuum's emptiness (to be discussed also). Obvious questions arise:

A. Can light ever get 'tired,' that is, ever slow down in a vacuum? Is it, has it forever been, and will it always be traveling at exactly 670,615,200 miles per hour? Now for 13.7+ billion years?

B. Lightspeed is measured <u>through</u> space. Space may also have (actually <u>has</u>) a measurable speed of its own expansion. Does this space expansion speed increase lightspeed in its same direction, or possibly decrease it in the opposite direction? Why or why not?

C. Light travels from point A at time <u>a</u> to point B at time <u>b</u>. But in the interim, points A and B may have moved, relatively to one another. Say they do so move, <u>but</u> they also <u>remain equidistant</u>. Light then will have taken a curved, therefore longer actual path between the two than geometrically exists. So, at constant (light) speed over a longer distance, light must take <u>more</u> time than <u>a</u> minus <u>b</u>. How is it possible if lightspeed is constant and A and B remain equidistant?

D. Given light from a galaxy some 3,500 light-years away. Say even 35,000 light-years away…and that's a great deal of distance. And if there are (actually there <u>are</u>) lots of gravitational entities, say, arbitrarily ±1,000 stars and ±100 black holes in 'the way,' so directly in this same light's path, said light will be <u>lensed</u> thereby…say (as in the above example) ±1,100 times. Lensing requires/ creates changes in direction(s), which increases actual, geometrical distance. Again, as in number C above, at constant lightspeed, said light should require more time in its actual journey. Is this a problem? If not, why not?

1. Assume in 2025, the United States and/or China put up a very sophisticated satellite in deep space and in a fixed position relative to our Sun. Said satellite absorbs sunlight and reorganizes it as a powerful laser and focuses it on object A, exactly <u>four</u> light-years away. In exactly four years, upon the laser's arrival at A, the satellite rotates the laser, still operating at full power, exactly 90° in, say, exactly one second… to object B, which is exactly <u>three</u> light-years away. When object B 'sees' that light in three years, it should also be theoretically invisible from object A, which (using the Pythagorean Theorem) is <u>five</u> light-years away at that moment. Has anything measurable, other than distance alone, transpired between A and B, five light-years apart, when laser light from the source reaches B only three light-years away? So then 'nothing' when/as this sun-laser sweeps across five years of space-time in a matter of a second? Has 'something' occurred at many, many times the speed of light? Will any observer 'observe' the transition? Ever? Or…

2. Given a very powerful lighthouse on a rock in the middle of an ocean or a locater beacon on a satellite in space. Assume it rotates rapidly at a full 360° once a second. At some point/distance out from the source/lighthouse, said light or signal will sweep out a circumference of exactly 186,282 miles. Such a radius would be 186,282/3.1416/2 or roughly 29,648 miles. Beyond this radius, would not observed light or signal be traveling in excess of 186,282 miles

per second? If not, why not? Would not an observer at 30,000+ miles out observe something, as in a phenomenon moving or rotating past faster than lightspeed c? If not, why not?

One could go on and on, but the subjects <u>are</u> fascinating. Thank you, Dr. Einstein. (Refer to chapter 7, "E = mc^2 (Dr. Einstein)" and chapter 15, "The Absence of Anything.")
Light and lightspeed will be oft-visited topics herein.

Additional Notes on Lightspeed

E. It is accepted scientific doctrine that bosons (force-carrying, messenger particles), to include photons (light), gluons, and W and Z particles ('nuclear' forces) and Higgs bosons recently demonstrated (at CERN) but hardly confirmed (to transmit mass), and gravitons not yet demonstrated nor confirmed (to transmit gravity), all bosons can theoretically stack up <u>upon</u> one another without any interaction(s) at all <u>with</u> one another. However, if the Higgs does impart mass, won't more of them in one place create more mass? Similarly for gravitons, when identified, won't more of them in the same place (<u>and</u> at the same time...same for all bosons) create more gravity? So then consider the <u>photon</u>. Do more of them in the same place and at the same time(s) simply/only create more (so brighter) light? Is there actually, absolutely <u>no interaction</u> between bosons under any circumstances over anytime or distance? If they cannot physically stack-up upon/with each other, do their energetic properties compound anyway? Somehow?

F. If a light beam from a very distant quasar has taken ten-plus billion light-years to reach Earth, its photons had to have been released in Earth's direction roughly 3.7 billion years after the Big Bang itself. That means said photons were released/sent (and for that matter still are), but the underlined original ones were sent almost one billion years prior to the existence of Earth (±4.5 billion years ago). It is hard to imagine that over this time and through such vastness of space, the original photons have always found themselves at a "perfect speed" at exactly 299,792,458 meters per second...again exactly and for 'ever' (say for the interim ten billion years, anyway). Light slows down a bit in Earth's atmosphere, so its last (trivial) transit to an earthbound telescope comes at 1/1.0003 c, or roughly 299,700 kps. One must assume, and it has been confirmed, that massive collections of interstellar/intergalactic gases and dust and presumably unknown other 'debris' exist...and must function much like cosmic atmospheres. So light's net speed from point A (the Quasar) to point B (the Earth), and wherever these points may be now relative to each other, cannot possibly have been measurable at exactly 299,792,458 mps...can it?

G. If light slows down in a medium other than in a perfect vacuum, and its speed through transparent glass (any encountered enroute?) is much slower, ±1/1.5c or ±200,000 kps, then how can it, so then how can cosmic distances be confirmed for sure? Recall $V = d/t$ and $d = Vt$ so $t = d/v$. If V changes, d and/or t must also. Yes. (Refer to chapter 7, "$E = mc^2$ (Dr. Einstein).")

H. Light lenses around gravitational objects...any and all of them. Given the same ten-billion-year transit/trip above, as in Note F, a lot of lensing must have occurred, therefore. Obviously, a lot of gravitational objects were encountered. Such lensing causes deviation(s) in the otherwise straight-line paths of photons. They deviate because the spaces through which they travel are warped...over and again by gravity. Thank you, Dr. Einstein. However, at time 1 at point A, the quasar (above) exists, for that moment, at space-time point 1A, exactly when its original (so let's assume first ignition) photons are released at exactly lightspeed = c. Again, they are/have now been received at time 2 at point B where Earth is now, so at space-time point 2B. So taking space-time point 2B here, now on Earth, and tracking/ plotting the quasars space-times 1A and then 2A, these actual geometric distances will be unequal. If the universe is expanding as noted and for the sake of simplicity, assume the quasar above is receding from Earth in a straight line (so on/from the other side of the universe), there would be no proper (lateral) motion. If Hubble's expansion constant is true that relative distances and speeds of objects in space increase mathematically over time, then the quasar, originally at universal space-time point 1A, some 10 billion years ago, (its ignition) is now much more distant now at universal space-time point 2A, now 13.7 billion years, not 'just' 10 billion years, after the Big Bang. And therefore, connecting these two space-time points with straight lines, either from Earth's space-time points 1B or 2B, cannot possibly equal the actual transits or distances traveled by of the quasar's photons. Contradictions:

1. Via lensing, the quasar's photons have also traveled much more deviated (i.e. nonstraight) paths and therefore more actual distances as a result. And...

2. Via exposures to interstellar/intergalactic atmospheres, they may not have been able to travel

at exactly net c…so less. So, are quasars really as far away as they appear? And…

3. Ignoring (general relativity does) the fact that the speed of the quasar's recession (assuming the Earth's position to be stationary) does not affect the quasar-photon's speed, then…

4. How can the same photos all arrive at Earth at this its space-time 2B (now) at the same time(s) over such very different, actual distances? All noting, "ten billion light-years only"?

5. Do photons not ever "bump into" one another? Are they absolutely immune to any "self-interactions" at all? Do they pass through the "blackness" of deep space with such a scarcity of other photons at the same unimpeded speed as they do when rounding or passing through another active galaxy with such a massive surplus of other photons? Are they absolutely immune to any "bumps," so to any self-interferences at all?? Again, they do bump into electrons in our (any?) atmosphere and do more so and more often (so transit more slowly) through glass or through water, etc. So…

6. Is c always a demonstrated, actual net lightspeed for all photons in any circumstance? Are measurements over vast times and incredible distances always accurate? Never affected by c as may be affected by either:
   a. extra, actual distance(s) traveled, or
   b. some refractive index while in transit, or
   c. the expansion of space itself.

7. Note: 6a and 6c above would require c plus some velocity = an impossibility? 6b above would define

c <u>less</u> some velocity = a possibility, probably a "probability." Therefore…

8. Actually, how much of this actually matters unless there is some mechanism or some condition to explain events faster than c…possibly instantaneous events? Slower is understandable, but again, faster? So then, if evident, must "instantaneity" exceed lightspeed?

But is it, instantaneity, so 0 time, permitted at all?? It is not if our mathematics, as currently understood, is correct. A contradiction appears to exist, nonetheless. (Refer to chapter 19, "Nothings.")

Conclusion: The <u>actual</u> performance of electromagnetic waves (of light in particular) may or may not agree exactly with its theoretical (read: mathematical) description. Further discussion will be presented in *Fizzicks 201*.

Chapter 9

Mario's Champagne

Relatively speaking...about motion...and to put "rest" to rest.

Endless nonstationary examples exist. This is one.

Assume/imagine the "2021 Gran Tour de Ecuador." An auto road race, with its start/finish in Quito (the capital). The race has been won by an Italian driver, Mario, in his custom Ferrari. Mario pulls into his pit lane for congratulations and is handed a celebratory bottle of champagne and thus departs upon his victory lap. Mario uncorks the bottle and shakes it and sprays his adoring fans along his way. All is customary at the end of a good auto race of most all types...yes?

Since champagne is primarily flavored and carbonated water, so a collection of H_2Os, one can focus upon just one water molecule = one H_2O bit, and to its relative motions therefore...by the numbers of such motions, beginning with the smallest and the 'closest in'...as follow:

1. Mario's driving hand is steady, but his free hand is waving the champagne bottle to maximize the coverage of its contents. So our solo H_2O has unique directional movement components at any instant relative to Mario and his arm, his hand...and his celebration.

2. All the while, our H_2O molecule is being expelled out of the bottle by gas pressure (CO_2) in whatever direction the bottle is momentarily pointed…at approximately ten feet per second, relative to the bottle.

3. Mario is driving at ± ten miles per hour on his victory lap, so our H_2O molecule has a third overall direction, same as the Ferrari and relative to the course/lap, which is upon the road surface, and which is at rest (nonmoving) upon the Earth's surface. However…

4. The Earth is <u>not</u> at all at rest. It and therefore Mario's road surface is rotating to the east at ±25,000 miles (approximate circumference of the Earth at the equator, where Quito is) every twenty-four hours. So then $25/24 \times 10^3 = \pm1,042$ miles an hour. So depending upon the point of Mario's (roughly circular) victory lap, our H_2O assumes another motion somewhere between 1,052 (when heading east) to 1,032 mph when heading west…depending entirely upon the Ferrari's direction. This is now relative to an observer in a momentary position (i.e. not geosynchronous) above Quito, observing the Earth's axial rotation.

5. The Earth is also revolving/orbiting around the Sun every year, less six hours (hence requiring the addition of one calendar day in February every four years). Nonetheless, the Earth's orbit currently covers ±584 million miles each 'lap' and currently requires 8,760 plus 6 or 8,766 hours to do so, accordingly, resulting in $\pm584 \times 10^6 / 8766 = \pm66,620$ mph orbital speed. And this is expressed in a very complicated aspect to (a) Earth's rotation, (b) the time of year (the tour was in August, okay?), (c) the Ferrari's speed and

direction, etc., etc. And, all this would be observed relative to a stationary position above or within our solar system…of Earth observed again at 66,600 mph orbital speed. And any specific 'direction' of this speed would vary with every moment of Earth in its orbital position(s).

6. Our solar system is currently located some two-thirds of the way outside the center of our Milky Way galaxy. The exact center of the Milky Way is theorized to be (same as the centers of most known galaxies are assumed to be) a massive black hole, which in time might theoretically consume Earth, Mario, his car and our H_2O as well! Nonetheless, our solar system is currently spiraling inward in a very complicated aspect and at a very complicated (to compute) speed to that end. It is beyond the scope of this discourse (and of this author) to know this speed and/or specific direction, but our H_2O, Mario and planet Earth are subject to it in both a very profound aspect and velocity, relative to our cosmos, which is also subject to…

7. The Big Bang, which is again the currently most explainable and understandable starting point, both in space and in time, of our observable universe. All indications and observations show that all primary objects, including all galaxies, solar systems, and other suns and planets are expanding away from each other relative to a 'point' that would be, arguably, the original location of the creation of our observable universe. However and wherever and whenever (currently calculated to have been ±13.75 billion years ago) this event, our 'beginning,' happened, it obviously <u>did</u> <u>happen</u>, and since that time, we all,

including Earth and Mario/his Ferrari and our H_2O, have been a part of this basic expansion, which is likely beyond calculation. But, it would be relative somehow to observation from outside our universe… however possible.

Which begs relative observation number 8: Is our universe itself the only one, and is it stationary? And if so, or if not, this motion would be relative to what? Where would or could such an observer be? Hard to say let alone to imagine. But, given our universe is expanding, into what space is doing so? And is this space unmoving?

We can leave these last potential movements currently to the philosophers. By the same rationale but in reverse, we can consult the quantum physicists (that study the very smallest of known objects) to learn that atoms and particles (protons, neutrons, and electrons) are themselves spinning and are constantly in motion. And then, even the constituents of these particles, theorized to be "quarks" and/or "strings," are constantly moving and/or popping in and out of existence. And while in and possibly observed, they are in constant vibrational motion referred to as "quantum jitters" composing all sorts of internal motions and spins as well!

Now that we've thoroughly debunked any notion of anything at all being actually stationary, what has actually been discovered or learned? If/when considering the very smallest dimensions of, perhaps, strings or quarks as opposed to the very largest measurements of our universe and then of its possible location, it becomes obvious that the three spatial dimensions upon which we base most all our earthly observation(s) and current science, those which define (1) up and down, (2) left and right, and (3) backward and forward, are not enough. String Theory, mentioned above, requires at least nine spatial dimensions, for example, and is conceptually as difficult to explain as the actual speed and direction of our universal expansion might be. And...

Enter Mr. Einstein, whose single greatest contribution was to add time as being necessarily the key point of reference to our 'known' dimensions. Since the "relativity" of motion in this discourse assumes one exact moment of observable time only, the thoughts and considerations of our movements (i.e. of our nonstationary existence) are easily expressed. But, add time...then welcome to Mr. Einstein's "special" (so at constant velocity) and "general" (so at accelerated or decelerated velocities) concepts.

Suffice it to say that we, just as individual earthlings, are for sure in constant motion and are moving in at least five or six very different and complicated directions for sure and at any time...always "relatively speaking." When, where, and why we are would be the next most logical questions...stated and asked in order of increasing complexity. Nonetheless...

Nothing and no one is ever 'at rest,' this depending upon any observer's 'outside' viewpoint(s).

Chapter 10

The Tennis Ball Analogy

First visualize the simplest atomic structure that exists, however small it is…that of element number 1, hydrogen. It is the most abundant element in our universe and, again, the simplest. It exists as one proton, p, orbited by one electron, e. Its basic diagram has been drawn/presented as:

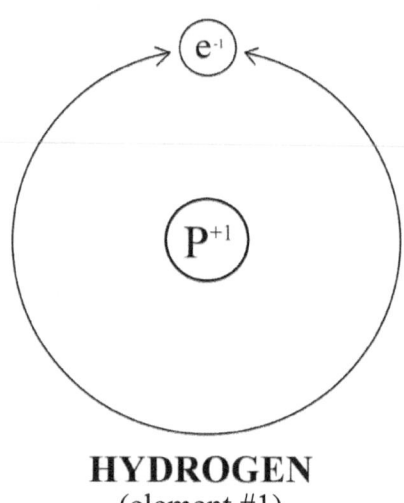

HYDROGEN
(element #1)

…and is terribly misleading.

Were one to magnify or otherwise enlarge an atom of hydrogen, so if somehow its nucleus, a single proton, were instead the size of a tennis ball, then its also single electron would be in orbit roughly one <u>kilometer</u> out/away! The above diagram, with the proton, P, drawn less than the size of a dime, would actually require some paper ten to fifteen feet wide on either side to fit in the electron's orbit then to be in proper proportion.

So then think of the <u>actual</u> proportions…in tennis-ball terms. A tennis ball is roughly 2.5 inches in diameter, so it has a radius of 1.25 inches. Figure a kilometer is .62136 of a mile, which is 63,360 inches. Then 63,360 × .62136 = 39,369+ inches in a kilometer. Add the 1.25-inch radius of a tennis ball and this virtual hydrogen's electron is ±39,371 inches distant, and that would be almost 15,750 additional tennis balls away from its nucleus in our analogy! And again, the first diagram (above) is totally misleading therefore.

Hard to imagine, but an entire hydrogen atom, including its electron's orbital space, no matter how big nor how small, is 99 percent-plus <u>empty</u> <u>space</u>!

Example 1: If one were to walk (at ±2-1/2 miles per hour = an average pace), it would require ±16 minutes to walk from the tennis ball proton to its electron…some 39,371 inches or ±3280 feet, so almost two-thirds of a mile away.

Example 2: Our home in Santa Fe sits at roughly 7,100 feet. Our local ski area begins at roughly a 10,200-foot base elevation. This 3,100-foot gain (to go skiing) is about the same as most major western ski areas' total verticals (when skiing)…and again a bit less than our enhanced electron's virtual separation. Again, either elemental or (virtually) tennis-ball-wise, this separation is proportionally immense.

It is hard to imagine. How can these primary atomic structures exist, let alone ever stay in sufficient communication and contact when <u>so</u> <u>far</u> <u>apart</u>? Given the 2-1/2-inch tennis-ball nucleus again, the electron is still but a 'point' of unknown measurement, and at an enormous distance...again, almost two-thirds of a mile away.

The binding energy (electromagnetic attraction) between a hydrogen's proton (+1) and its electron (-1) may be much like, but not necessarily calculated the same as, one electron volt on the atomic scale. Take the electron radius to nucleus distance ratio of 39,371 inches / 1.25 inches = 31,500 (times), and it is indeed remarkable their electromagnetic partnership exists at <u>any</u> level. However...

When (virtually) increasing the size of a single proton (again, a hydrogen nucleus) to the size of a tennis ball, one must also know to have to also increase its +1 electromagnetic attraction in proportion to and with its also (virtually) enlarged and (virtually) energized −1 electron. And this begs the analogy again between atomic organizations that are bound by electromagnetic attraction as compared to celestial organizations, demonstrated in/by solar systems and galaxies that are instead bound by gravity. (Refer to chapter 7 and notes, "$E = mc^2$ (Dr. Einstein)")

So, the same 'attractions,' whether electromagnetic or gravitational, apply nonetheless. Fortunately, protons and electrons <u>and</u> stars and planets, however differently energized, do organize themselves...and tennis balls, somewhere 'in between,' do not!

One might attempt to calculate exactly what the necessarily increased mass of an enlarged proton would have to be and also the increased mass of its electron (at 1/1836th) would have to be for its organization to shift from electromagnetic to gravitational. Some proportional consideration similar to that in chapter 17, "Dimensional Proportionality," must apply. TBD.

An electron's orbit(s) are multidirectional, so then necessarily spherical, not flat or planar, as is Earth's orbit around the Sun. Any one hydrogen electron's orbital circumference at any point around a tennis-ball-sized nucleus would be a bit less than four <u>miles</u>. Considering such an erratic orbit (insofar as it's said that an electron can be found at most any point around its nucleus at any time), the electron in our analogy would surely have to travel in excess of lightspeed... which is impossible. Recall, electrons do have mass, and it is a very good reason that actual dimensions of atoms are actually <u>so</u> small and not tennis ball sized, therefore. Actual atomic dimensions would allow for near lightspeed orbits.

Considering volumes: A tennis ball contains 1.953 (or r^3) × 3.1416 × 4/3 = 8.181 cubic inches. The volume of its assumed virtual electron 'cloud' (recall with a radius of 39,371 inches) would be in excess of two hundred <u>trillion</u> cubic inches! The approximate, volumetric ratio of substance (a 8.181-cubic-inch nucleus) to the electron's 'cloud' is roughly 1 to 312 billion...difficult to comprehend. But it would be the same ratio applicable to elemental hydrogen as well. Hence, this analogy.

How can and how do two hydrogens then bind to and unite with one oxygen to produce a molecular substance as dense as water? Carbon atoms, element number 6, hook up and when compressed form diamond crystals as well. It is difficult to explain the polished surface of a diamond (perhaps in a wedding ring), given all the space that must remain even after compression between/within its carbon (only) atoms. Why don't we fall through the floor, especially wooden (so 'loose') ones? Accordingly, how do <u>we</u> 'hang together'? We too are more space than substance. Some explanation exists via...

Electron one in hydrogen's first (only) and outer, or "valence" ring, is 'alone.' By nature's laws, again necessarily applied and not otherwise explained, valence ring number one wants two electrons to fill. So hydrogen, with only one electron, is very reactive and will gladly respond and bind to other elements to gain that second electron…and so to form many, many compounds or molecular structures as a result… H_2O included. But…

When element number 2, helium, is formed (via fusion of hydrogens…explained in chapter 12, "Particle Gender"), helium's valence ring number one gains its 'necessary' second electron and is 'full' thereby. Helium exists as a "noble gas" and is essentially <u>non</u>reactive with two electrons only in its (full) first ring. Next electron ring number two, expanding outward, wants eight electrons to fill. Element number 3, lithium, has three protons and three electrons, so two in the first (full) electron ring, but only one in its second, or valence ring that 'wants' seven more to fill. Lithium too is very reactive, as opposed to…

Element number 10, neon, which has two electrons in its first ring (so full) and eight in its second ring (also full). Neon, like helium, is a noble gas and is also nonreactive. The logic about numbers of electrons needed, versus the numbers actually included in electron valence rings, is very expressive of an element's availability or nonavailability to combine with others.

One can go on and on explaining the increasing complexity of heavier and heavier elements. Ring number 3 also fills with eight (more) electrons (so has the same requirement as ring number 2). Argon, element number 18, the next heaviest noble gas, has two electrons in ring number one, eight in number two and eight in number three—eighteen total and with all its rings full. Argon is also a nonreactive 'noble gas,' therefore. The logic continues.

Each time a ring adds an electron and then ultimately fills, it contracts so is drawn (or forced?) closer to its nucleus, and the entire atomic structure compresses…so it becomes more dense. The resulting tension is expressed as 'electron resistance' (a resistance to further compression). And as this tension increases and as atomic organizations become closer together, so then they become more condensed, this keeps us from falling through wooden floors, etc.

Heavier and heavier elements become more difficult to compress. Element number 92, Uranium 238, its heavy isotope in particular (another discussion), becomes <u>very</u> unstable when sufficiently compressed, for example, and the very complex mechanics of 'explosive nuclear fission' results, unfortunately…and is another subject for *Fizzicks 201*, perhaps.

Back to our tennis ball analogy: Beginning with element number 1, hydrogen, element number 2, helium, adds one proton and also two neutrons to its nucleus. Noticeably there exist no stable elements (so then isotopes only exist) with two only or with three only particles (baryons) in any nucleus. The count for elements goes from atomic weight one, hydrogen, to atomic weight four, helium, element number 2. And helium may be diagrammed as below:

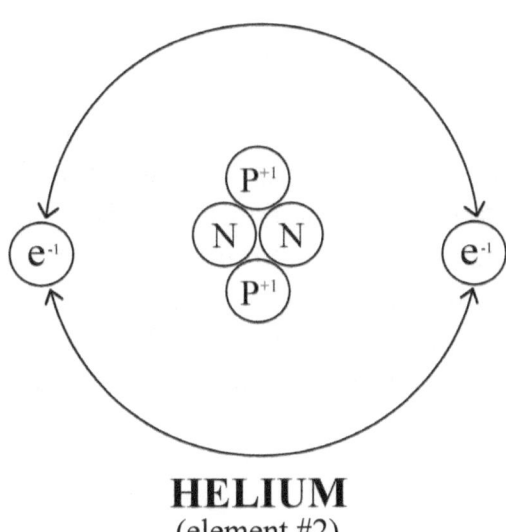

HELIUM
(element #2)

Again, given a multiple tennis ball analogy, this diagram is also misleading. Helium's two electrons also orbit with proportionally very distant paths. Helium, like hydrogen, must be at least 98 percent empty space. It is beyond the scope of this chapter to know if or when elements actually become less than, say, 90 percent empty? None to this author's knowledge would even approach. So, the elements and compounds that make up everything we know defy common logic as to their actual structure(s) that we can experience as solids or liquids…or even gases, if one thinks 'hard' enough. Much, much more space than substance exists in our world's 'stuff,' regardless of its simplicity (say, of hydrogen gas), or of its complexity (say, of the core of a nuclear bomb)…prior to its detonation!

The "natural laws" to which we are all subject, including the finite and fixed top speed of light and the specific electron requirements that exist to fill electron rings numbers 1, 2, 3, 4, 5, 6, and 7, and many others, are established and unchangeable. It's good to know that because…

If any of said laws were even slightly different or are ever violated, neither our solar system nor we, man, would exist as an entity or as a species. Actually, life itself would not exist as we know it. Laws in physics exist as extant and inviolate and apparently as 'intended.' Whose 'intentions' that may be involved is another consideration. However…

It's best to remain 'obedient.' Ours is a "law and order" universe. Perhaps *Homo sapiens* might best take some political and societal lessons therefrom? Might more progressivism at this time be ill advised? And might Dr. Sagan's observation of Earth as a "Pale Blue Dot" becomes all the more poignant?

Chapter 11

The Significance of "3's" (and of 1/3's)

Regardless of our many languages and the various names of numbers, our world's mathematics is 'ten -based.' Ours is a <u>decimal</u> system, easy to use. Move decimal points or add or subtract exponents and computations are greatly simplified. However…

Threes (and their reciprocal 1/3's), and all multiples thereof, all play very interesting, very arbitrary, and very pertinent and very present subroles. So much of our world is organized and presented "in triplicate." To explain:

1. Any number whose integers add up to a total that is divisible <u>by three</u>, is both itself a product of and is also divisible <u>by three</u>. If unclear: 3, 6 and 9 are obviously products and are divisible…and so also are 12 (1 + 2 = 3), and 15 (1 + 5 = 6 / 2 = 3), and 18 (1 + 8 = 9 / 3 = 3). And, so is and so are the following, for examples:
    - 21 = 2 + 1 = 3 / 3 = 1
    - 221 = 2 + 2 + 1 = 5…so, no. But…
    - 222 = 2 + 2 + 2 = 6 / 2 = 3, so, yes.
    - 2222 = (integer) 8 = no, but…
    - 2223 = (integer) 9 / 3 = 3, so yes. And try…

- $5112 = 5 + 1 + 1 + 2 = 9 / 3 = 3$, or
- $5,112,637 = $ (integer) $25 = $ no, but
- $5,112,636 = $ (integer) $24 = 2 + 4 = 6 / 2 = 3$, so yes.

...and so on...infinitely (so assumed!). Note that 3,258,946,872,692,451 is divisible by 3 because its integers add up to 81.

This very unique relationship holds...just like $3 \times 4 = 12$ (integer total 3) or $3 \times 5 = 15$ (integer total 6) or $3 \times 6 = 18$ (integer total 9) does...and so on ($3 \times 7 = 21$ or integer 3 again). It is the property of our ten-based mathematics that always works for 3's and when adding integers...curiously. Accordingly, $3 \times 5,115,234 = 15,345,702$, which integers add to 27.

Note here: Lightspeed, for example, in meters per second equals 299,792,458...whose integers add to 55—so not divisible by 3. But, subtract <u>one</u> meter only, and 299,792,457 integers total 54, and $5 + 4 = 9$, is so divisible by 3...if this matters...which it may! One (only) less meter per second would make the difference.

How this and other properties of 3's and of 1/3's do or do not apply per the following examples may be notable according to anyone's determination, rationale, or even intuition? Many relationships simply 'occur,' so they exist as in very defined and specific "triplicates," nonetheless. To explain further:

2. Given our arbitrary meter (its exact measure), lightspeed <u>almost</u> equals 300,000,000, so $\underline{3 \times 10^8}$ meters per second. Were our meter just a fraction shorter (as well it might be), or were our second just a fraction longer (discussed just below), then light's

speed would measure exactly at 3×10^8 meters per second...again, exactly and currently. But again, 300,000,000 less 299,792,458 is a difference of 207,452 meters per second. This fractional difference amounts to an error of .000692 (say 7 ten-thousands)...not much. However...

Earth's rotation appears to be slowing, ever so slightly, and this is thought to be due to the 'drag' created by the oceans' tides. One estimate put a day, so one rotation, for the dinosaurs (so mid-Cretaceous) at 23 hours, 59 minutes, 56 seconds...4 seconds quicker just a few million years ago. In another few million years, perhaps our day will last for 24 hours, 4 seconds? Nonetheless, our standard 'second' must remain inviolate... And so must our lightspeed...as expressed as almost 300,000,000 meters (also inviolate) per second...again, does this matter? Our days only will be increasingly longer, nonetheless, and clocks (and watches) will have to adjust accordingly! A 'day' may have to last for 86,404 seconds 'some day'?

3. Our universe exists in three only, distinct spatial dimensions as demonstrated via (1) height, (2) width, and (3) depth. At any one point in space, three only lines (or 'rods') may be drawn (or placed) with 90° angles between each of them. The early Greek mathematicians and 'geometrists' figured this out to be true. More recently, Dr. Einstein added a fourth dimension of time...but in a "relative" sense only. Time is 'temporal' and therefore nonspatial. Three only spatial dimensions are applicable and measurable in our currently observable universe at any moment in time, therefore.

4. Triangles exist as the simplest solid in two dimensions, just as three-sided pyramids do in three dimensions. Note: This is evident upon the addition of a fourth side as the base. Four-sided structures exist as squares or rectangles in two dimensions. Then add a base and a top, a total of six sides, as cubes…and so on. No two-sided solids or structures exist. All solid structures must have at least <u>three</u> sides.

5. Take/make a line (any length), fix its center, and rotate it. Its ends will scribe out a closed arc, which is a circle. The ratio of the length of any such line, or diameter, to the length of any such circle, or circumference, is always 3.1614 or Greek letter pi…as confirmed by Archimedes, a Greek mathematician, roughly 250 years prior to the birth of Christ. So the simple <u>one-dimensional</u> formula for the <u>one-dimensional</u> length of any <u>one-dimensional</u> circumference is any <u>one-dimensional</u> diameter times pi. This is elegant and obvious…and involves exponent <u>one</u> only.

6. Given any such circumference, again any one-dimensional circle, to find its <u>area</u>, requires <u>two dimensions,</u> <u>both</u> height <u>and</u> width. Such a formula to express <u>two dimensions</u> requires that any diameter be halved to become a radius, that is <u>squared</u> (multiplied times itself), so taken to the <u>second power</u> times pi. So to calculate any two dimensional area requires (in effect) a squaring…which is a power of <u>two</u>… exponent <u>two</u>.

7. To express any volume then requires <u>three dimensions,</u> so to include depth as well. Take any line or rod (as above), fix its center, and spin it in <u>all</u> directions, and its ends will scribe out a sphere,

which is a three-dimensional circle. One may assume then that some mathematical power of three may be involved…and so it is. The formula for the volume of any sphere uses the same measure of any radius times itself, and then times itself again…so cubed, and therefore to the power of (exponent) three, then also times 4/3 pi. Curiously, it is $3/3^{rds}$ plus $1/3^{rd}$ = $4/3^{rds}$ of pi $\times$ r^3 that applies…always in factors of one-thirds.

8. The formulas, as based upon the specific dimensions involved, so then utilize the same mathematical powers as do the dimensions involved. This is an expression of mathematical 'poetry,' perhaps? Form and function appear to behave in tandem:

   a. One-dimensional circumference is expressed by any one-dimensional diameter (exponent one) times pi: $C = d \times pi$.

   b. Two-dimensional area is expressed by any radius (half diameter) squared (exponent two) times pi: $A = r^2 \times pi$.

   c. Three-dimensional volume (i.e. of a sphere) is expressed by any radius cubed (exponent three), times by 4/3 pi: $V = r^3 \times 4/3$ pi.

So then, to go back to the two-dimensional surface area of a sphere, the computation converts quite elegantly back into any diameter squared times its circumference: Area of a sphere = d^2 C…again two dimensions produced by a power of two. And the mathematical 'poetry' continues.

9. Note that the same considerations apply to any and all symmetrical solid(s). The perimeter of a one-dimensional square is simply four times the length

of any side. So then, the <u>two</u>-dimensional area of any square is the length of any side times itself (so 'squared'). And then, the <u>three</u>-dimensional volume of any (also, <u>three</u> dimensional) cube is the length of any side times itself, and then times itself again…so 'cubed.' And the 'poetry' continues.

These forms (symmetrical solids) express their dimensional parameters (lengths, areas, and volumes) via the same exponential powers (one, two and three). And, three only special dimensions exist, accordingly.

10. One might then logically inquire if the power of four (x^4) is involved in any computations that might involve space (dimensions 1, 2 and 3) <u>and</u> time (dimension 4). Space-time computations involve the mathematics (primarily Dr. Einstein's) of relativity…and are well beyond the (promised) limits of this writing. But yes, exponent(s) 4, so to the 4th powers, apparently can become evident in such advanced mathematics.

11. Note that a circle, such as a compass, could be measured or might be calibrated in/by tens or by tenths. So such a circle, which might be a compass (or a clock, as discussed below), might be organized to express 100 or 1000 degrees of rotation. Right angles of such a ten-based circle would then measure 25° or 250° rather than 90°. But compasses and clocks are <u>not</u> so expressed. Compasses (their circles) are organized by/in 360°, and clocks by/in 12 hours (notably 3600 seconds per hour), both with <u>three-based</u> increments. Why, one might ask, given that our mathematics is ten-based?

12. Clocks are another circular measure of <u>time</u> expressed in hours (24 per day/night and all in factors of 3), minutes (60, ditto), and seconds (3600 per hour, also). Time measurement(s), if expressed, (however re-calculated) in tens or tenths, would contradict our <u>three</u> dimensions of space, again as would degrees of a circle, in tens or tenths. Note: 10/3 = 3.33333... n^{th} = a never-ending calculation and never being exact...more is discussed on this below. However:

a. Note also that 360°/4 = 90° right angles, so compasses correspond to our cardinal directions: east, south, west, and back to the north—as expressed in the clockwise direction of area(s), as measured in three-based increments.

b. Times (clock-face) 12 hours/4 = the same ¼ hours expressed also in factors of 3:

   i. 90 ° = east = 15 minutes = 1/4 of daylight, etc.

   ii. 180° = south = 30 minutes = 2/4 of daylight

   iii. 270° = west = 45 minutes or 3/4 of daylight...etc., and

   iv. 360° = north = 60 minutes or one day or night.

   v. All measurements are of/in factors of <u>threes</u> again but also <u>in</u> <u>four</u> <u>aspects</u> <u>to</u> <u>confirm</u> <u>space</u>-<u>time</u>. So clocks and compasses are both unique in relating both <u>space</u> and so <u>direction</u> to <u>time</u>.

Of note: Is there a connection here? Space-time adds the fourth dimension—again, time. So circles and compasses are expressed accordingly: North-south and/or east-west are expressed by vertical and horizontal lines (diameters) that intersect perpendicularity (90°)...so then in the four cardinal directions. It is the same for clocks. Twelve (noon or midnight) would connect to six (so "half-past") vertically (same as north-south) and three o'clock (a.m. or p.m.) would relate in the same way horizontally to nine o'clock (a.m. or p.m.) as east-west. It's a logical 'marriage' of <u>three</u>-dimensional spatial measurements combined with <u>one</u>-dimensional time to express <u>four</u>-dimensional space-time, which corresponds to our actual existence. But all base measurements are managed by increments (degrees and/or seconds) that are expressed in multiples of <u>three</u>...not of ten(s)

13. Again, a triangle with <u>three</u> <u>only</u> sides is the simplest 'solid,' so the simplest <u>two</u>-dimensional form. And its <u>three</u> only interior angles equal $\underline{3} \times 60°$ to total 180° on a flat, so then upon a <u>two</u>-dimensional surface. On a <u>three</u>-dimensional surface, so then upon a perfect sphere, and using any circumference, the internal angles of a right triangle equal $\underline{3} \times 90°$ or total 270°. All such angles are again expressed in dimensional multiples of <u>threes</u>, regardless. Back a moment now to "just the numbers."

14. The first whole number "one" is unique...again as being the first whole measurement or 'quantification' of anything that is measurable. Anything less (except zero—another consideration for another time) than one and greater than zero is fractional. One, as if by 'design,' is the only number other than zero, when multiplied by itself, remains unchanged. Note: $1^1 = 1^2 = 1^3 = 1^4 = 1^{10} = 1$. One is also 'prime,' insofar as it may be divided evenly only by one or by itself—conveniently being also one! Likewise, the square root of one is one—also the only number other than

zero to behave accordingly. So any number to the first power is itself...$1^1 = 1$ as above, so then $10^1 = 10$, $11^1 = 11$, $12^1 = 12$...and so on. Note: Then accordingly, any number (including zero) taken to the zero power equals one.

15. (So integers $1 + 1$, coincidentally?) Two is also unique, also 'prime'—the only 'even' (so divisible by 2) prime number. Two may be squared (taken <u>to</u> the power of <u>two</u>), so 2^2 or simply added to itself. So 2×2 or $2 + 2$ both equal 4...a characteristic of its own and of none other(s). The reciprocal of 2 is then $1/2$. Note then:

   a. $1/2 + 1/2 = 2/2 = 1$, and
   b. $1/2 = .5 + .5 = 1$. Two and its reciprocal ½ are rational number(s) therefore. But...
   c. Note that the square root of 2, unlike the square root of one, is not...so the root of 2 is an <u>irrational</u> number. Its calculation, like that of pi and the reciprocal of 3, noted below, can and do go on forever. Irrational numbers' calculations seem to have no endpoints, so no exact answer(s).

(Again integers $1 + 2 = 3$) Three will be discussed below after 4 and 5 because...

16. (Integers $1 + 3$) Four is rather 'normal'...finally, since 1 and 2 (above) and 3 (below) are not. Four is not prime (so <u>is</u> divisible by two, as it and all larger, even numbers are). The reciprocal of 4 is $1/4$. Note then:

   a. $1/4 + 1/4 + 1/4 + 1/4 = 4/4 = 1$, and...
   b. $1/4 = .25 + .25 + .25 + .25 = 1$, exactly.

Four, and its reciprocal, are rational. Expressly…

    c.   .250000…n equals .25 <u>exactly</u>, and 4 × .25 = 1 <u>exactly</u>.

17. (So integers 1 + 4 again—and I <u>promise</u> I did not 'plan' this!) Five is also normal and prime. It is also the primary subfactor of our ten-based decimal system. The reciprocal of 5 is 1/5. Note then:
    a.   1/5 + 1/5 + 1/5 + 1/5 + 1/5 = 5/5 = 1, and
    b.   1/5 = .2 + .2 + .2 + .2 + .2 = 1 exactly.

Five and its reciprocal are rational.

    c.   1/5 = .200000…n so .2 <u>exactly</u>, and times 5, again = 1 exactly.

18. (So integers 1 + 5 = 6/2 = 3!). Three, the 'revolutionary' number, is prime, but again is abnormal. It is the basis (for reasons expressed in notations, numbers 11 and 12, above) for 360° circles and twelve-hour clocks, both expressed also in 60 <u>time</u> seconds or in 60 <u>arc</u> seconds. Some of its other characteristics/observations will be noted below. <u>But,</u> the reciprocal of 3 is 1/3. Note then, <u>however</u>:
    a.   1/3 + 1/3 + 1/3 = 3/3 = 1, <u>but</u>
    b.   1/3 = .333…n. So .333 + .333 + .333 = .999…n— and does <u>not</u> equal one. Note then also…

Three's arithmetical expression (divisions of/in 3's or in $1/3^{rds}$) does not agree exactly with its mathematical expression. As in:

    a. $1 \times 3 = 3 \times 2 = 6$, and $1/6 = .166666\ldots n$ and is not rational.

    b. $3 \times 3 = 9$, and $1/9 = .111111\ldots n = $ not rational (but repetitive like $.3333\ldots$ a separate consideration).

    c. $3 \times 4 = 12$, and $1/12 = .083333\ldots n = $ not rational.

And so on. This contradictory expression for all multiples of 3 (and reciprocals) goes on infinitely$\ldots 1/3^n \times 3 \neq$ one! Such a possible mystery may be resolved, perhaps by...

19. The 'magical' or mathematical square? Referenced herein is *Math for Mystics*, by Renna Shesso...with the following:

    a. Numbers one through four cannot properly form a 2×2 mathematical or magical square:

1	2
3	4

2	3
4	1

or

1	4
3	2

1	3
2	4

Note: That however numbered, the additions of these integers, either horizontally or vertically, all produce varied and/or different totals. But...

b. By taking the number 5 as central, and using numbers 1 through 9, and then constructing a 3 × 3 magical square, the same kinds of totals equal:

4	9	2
3	5	7
8	1	6

Note: the mathematical repetitions, as expressed below:

c. Total integers above 1–9 add up to 45 (also divisible by 3). When divided by 3 (referenced by 3 × 3 from any/all sides) = 15. And…

d. Any/all 3 sequences (there are a total of 8), either horizontally or vertically or diagonally, also add up to 15. Any way it is added…the total is always 15…and all in factors of 3. But…

e. If 5, the center digit, is doubled (so to 10 = our decimal representative), all totals, either sides of the 5, so 4 + 6 or 9 + 1 or 2 + 8 or 3 + 7, etc. do equate also to total 10. So…

f. Threes, or factors thereof, can also express themselves logically in/or within a ten-based decimal system. And…

g. (As in number 11 above), threes can be expressed in 4^{ths}, so then the four 'cardinal' (two-dimensional) circular directions agree, or…

h. (As in number 12 above), they can be expressed in our time (so in four-dimensional space-time) directions as well. What a wonderful and flexible and expressive value 3 <u>is</u> and its multiples <u>are</u>!!

i. dimension one, as $3^1 = 3$ obviously. And…

ii. dimension two, as in $3^2 = 9$ to express areas, etc. And…

iii. dimension three, as in $3^3 = 27$ expresses volumes or cubic, <u>three</u>-dimensional objects/measurements as well. And…

iv. (coincidentally) relate to base $3 \times 4 = 12$, our time…so circles (compasses and clocks) also reflect space-time. And…

v. (again) $5 \times 2 = $ base 10 in which our mathematics is also expressed via a <u>3×3</u> magic square.

Impressive, for sure. And, in addition…

20. Our very existence must be subject to <u>three</u> very specific and primary velocities…those being:

a. $c = $ light speed or 3×10^5 kps, and

b. $g = $ gravity speed, (same as c), plus…

c. t must equal time speed (same again?) How to enumerate?? TBD .

Note again: These first <u>three</u> expressions of our primary velocities must somehow relate directly to our <u>three</u> primary spatial dimensions: height, width, and depth. TBD.

Note also and again, all <u>three</u> velocities above must also, each equal roughly <u>3</u> $\times 10^5$ kilometers per second. (Damn that slightly too-long meter again! And, how is it best to measure the 'speed of time'?) TBD.

21. Any spatial measurement must decide between and therefore involve one of two of the <u>three</u> options, so then between <u>six</u> <u>actual</u> directions, be they: up or down, right or left, or forward or backward... total six. But the choices in either direction provide two options only...so 6 (directions) divided by 2 (choices) equal 3 (only) results! One must go up <u>or</u> down, right <u>or</u> left, or forward <u>or</u> backward...again, <u>3</u> choices only.

22. Any and all atomic structures (so elements) consist of <u>3</u> <u>only</u> primary or basic particles, so then of <u>3</u> <u>only</u> elementary fermions:
    a. protons
    b. neutrons and,
    c. electrons

...and all are "matter" particles that (by definition) "take-up spaces"—spaces that coincidentally have <u>3</u> dimensions. (Note only here that the exact dimension of an electron remains unconfirmed...so far). And, curiously...

23. <u>Three</u> <u>only</u> atomic structures of atomic weights 2 or 3, all isotopes, exist on line one of the Periodic Chart of Elements. So, in between element one, hydrogen (atomic weight one and in position one) and element two, helium (atomic weight four and in position 18) exist only:
    a. deuterium, heavy hydrogen of atomic weight two, and
    b. tritium, <u>very</u> heavy hydrogen of atomic weight three, and
    c. light helium (He_3) also of atomic weight three.

Note: These <u>three</u> rather elusive isotopes of elements one and two are rather 'temporary' and involved specifically in primary nuclear fusion (another discussion for *Fizzicks 201*). But, of the 16 vacant spaces on line one of the periodic Chart, <u>three</u> <u>only</u> 'temporary' isotopes exist—not exactly as members, but as the unique and the only isotopic atomic structures that might be included on line one, nonetheless.

24. And of particular notice: The primary, confirmed, and measured <u>nuclear</u> particles, so "nucleons," are <u>protons</u> and <u>neutrons</u> only. Both consist of <u>3</u> <u>each</u> "quarks" (thanks to Murray Gell-Mann). Each quark expresses/contains fractional electric charges of $1/3^{rds}$ each, as follows:
   a. Protons consist of <u>3</u> quarks:
      i Two "up" with charges of $+2/3$ each, and…
      ii One "down" with a charge of $-1/3$, thus…
      iii The resulting <u>proton</u> charge is $2/3 + 2/3 - 1/3 = 3/3 = +1$—again expressed in $1/3$ multiples.
   b. Neutrons also consist of <u>3</u> quarks:
      i. One "up" with its single $+2/3$ charge, and
      ii. Two "down" with $-1/3$ charges each, so… $2/3 - 2/3 =$ zero.
      iii. A neutron's charges cancel to zero. <u>Neutrons</u> are <u>neutral</u> as would be expected.

So the very basic constructions of all "nucleons" are 3 quarks with $1/3^{rds}$ electrical charges each…<u>all 3's</u> or reciprocal(s) thereof. No <u>twos</u> (or less) and no <u>fours</u> (or more)…only <u>threes</u> or <u>one-thirds</u> are involved.

25. Beyond consideration of quarks contained in nucleons only, there are two more, so a <u>total</u> of <u>3</u> families or generations of elementary "fermions" or matter particles:

    1st. Up and down quarks with +2/3 and -1/3 charges as already noted just above represent generation one.

    2nd. Charm and strange quarks, of heavier masses, but the same +2/3 and -1/3 charges exist as generation two. And...

    3rd. Top and bottom quarks, the most massive, but again, expressing the same +2/3 and -1/3 charges exist as generation three...and none others.

26. In fact, and in addition, <u>only three</u> separate electrical charges are yet known to exist for any/all particles, be they fermions, leptons, bosons, or baryons... those being charges of -1, or zero or +1 = total <u>three</u> electrical charges:
    a. -1 carried by electrons, muons, and taus (3)
    b. zero carried by neutrons, photons and gluons (3), and
    c. +1 carried by protons and one each of <u>3</u> weak-gauge bosons (subnuclear particles)
       i. w+,
       ii. w-, or
       iii. z° bosons a...being subatomic particles that express the weak-nuclear force. TBD.

Note: Positive 1 charges are also carried by <u>3 each</u> pions, as noted below. No charges of +2 or 'larger' or -2 or 'smaller' seem to exist. So it's again, <u>three</u> whole charges only:

    a.   plus one

    b.   none (so neutral), or

    c.   negative one...and none others...yet, again other than the fractional charges of quarks... notably also in increments of $1/3^{rds}$ <u>only</u>.

27. All particles to date also exhibit 'spin' or better, <u>rates</u> of spin...either zero (none), or one-half, or 1...again <u>3 spins</u> <u>only</u> seem to apply to date.

    a.   Spin zero for <u>3</u> <u>only</u> pions (noted above). This ignores the Higgs boson, an oddity to date and noted below after item number 29 below. TBD.

    b.   Spin 1/2 for <u>3</u> groups of <u>3 only</u> quarks and <u>3 only</u> leptons as noted. And...

    c.   Spin 1 for bosons, so for force carrying particles, two identified and one theorized, total <u>3</u>:

       i.   photons

      ii.   gluons, and

     iii.   gravitons (theorized)

So <u>three</u> spins only are recognized to date: zero, 1/2, or 1. The graviton, theorized but not yet confirmed, is theorized to exhibit a spin 2, solely of its own. There may also be some theoretical possibility of fractional spins, so of 2/3 or 1/3 ($1/3^{rds}$ again)...as <u>may</u> be discovered and confirmed one day? TBD. But, as of now, <u>three</u> spins only apply. Also...

28. Quarks demonstrate <u>three</u> <u>only</u> 'colors' as well... be they arbitrarily itemized as red, green, or blue.

Of note then: <u>Three-only</u> primary colors (above) are said to exist in (white) light. For nonscientists, it is difficult to explain then "yellow" or "orange." It seems also that <u>three-only</u> primary <u>pigment</u> colors are then red, yellow, and blue. Note: red and yellow produce orange, as yellow and blue do green, etc. It is the differentiation between 'white-light' and pigment color. <u>Three</u> primaries exist either way…as do <u>three</u> <u>only</u> 'colors' for quarks.

29. Truly, <u>three</u> <u>only</u> distinct and observed and confirmed "kinds" of force-energies exist in nature:
    a. electromagnetic.
    b. nuclear (assuming herein that both "strong" and "weak" versions thereof have a common purpose). And…
    c. gravitational.

Note here: The future may describe a 4th (or 5th), a "life force," perhaps? To be discussed in *Fizzicks 201*.

30. There exist <u>three</u> <u>only</u> families or 'generations' of all known particles to date: Quarks, with +2/3 and -1/3 charges are noted above. So then also:
    a. Leptons…3 groups exist,
        i. electrons and electron neutrinos
        ii. muons and muon neutrinos
        iii. taus and tau neutrinos
    b. Mesons…<u>three</u> <u>only</u>:
        i. positive pion
        ii. neutral pion, and
        iii. negative pion

    c. Bosons with mass, <u>three-only</u> weak-gauge nuclear force particles:
    1) W+
    2) W-, and
    3) Z°

Note here: The recently observed (at CERN) Higgs boson <u>is</u> an oddity and indicates a very high 100–200 GeV mass. Possibility here: To expect another two high-mass bosons? TBD! The Higgs needs two more partners to conform!

    d. Zero-mass bosons, again <u>three only</u> (one theorized):
    i. photons
    ii. gluons, and
    iii. gravitons (theorized)

31. And an obvious last cosmological observation: Stars, when their cores fuse to iron, seem to collapse structurally into three general, two known, and one theorized entities:
    a. white or dark dwarfs
    b. neutron stars, or
    c. black holes (currently theorized)
32. Matter. Of note, within the <u>three-only</u> spatial dimensions to which we are immediately subject exists <u>matter</u>. It, matter, so all the atomic structures that we know and of which we all exist, be they elements or isotopes or organic or inorganic compounds, also exist in <u>three-only</u> natural states to which we are immediately subject:
    a. solids

b. liquids, or

c. gasses

Note A: Solids, when sufficiently heated, can behave like liquids…think of volcanic lava flows or molten glassworks. So also can matter, when extremely superheated and subjected to extreme pressures, convert into plasma states. However…

Note B: Neither of the altered states above, including solids behaving as liquids and/or extreme plasma states, exist by themselves within the temperature and conditions that permit life. Accordingly…

Note C: Life on Earth exists within a relatively narrow and confined temperature range, currently as measured between Ø° to 100° centigrade, the corresponding freezing point Ø° and the boiling point 100° of water. So then, via deduction, liquid water becomes a critical factor for the existence and continued sustenance of life itself <u>as we know it</u>. More extreme examples of living organisms in -1°C (Centigrade) or less or 101°C or higher environments exist but have adapted as variants.

Note D: Within the confines of Ø° and 100°C (roughly 32° and 212°F (Fahrenheit) at sea level), all matter exists naturally as either a solid, liquid, or gas…<u>three-only</u> states… as above.

33. Water…not an element, but an inorganic compound of two hydrogens and one oxygen…of note (coincidentally?) of <u>three-only</u> atomic structures. Water is unique in its ability to exist in <u>all three</u> states of matter, as above, be they:

a. a solid, so as ice at temperatures of Ø°C and below.

    b.   a liquid…again its 'natural' state that permits life, between, the $\emptyset°$ and 100°C "sweet spot" as above, or…

    c.   a gas…so steam, as vaporized at temperatures above 100°C.

Note A: Water will also evaporate as cold <u>vapor</u>, also a gaseous state, at temperatures within life's sweet spot. Cold vapor, like steam, will tend to condense back into liquid water, its preferred state, again within $\emptyset°$ and 100°C. But water appears to be the only, common substance extant in <u>all</u> <u>three</u> matter states under normal, life-permitting temperatures.

Note B: Other gasses, notably elemental nitrogen and the compound carbon dioxide (CO_2) can/do exist as liquids (nitrogen) and solids or "dry ice" (CO_2) at extreme temperatures and pressures, but again, not naturally and not within Earth's sweet spot…and not without assistance.

34. Mercury. This element #80, Hg, is known to exist within Earth's sweet spot as one of two, elemental <u>liquids,</u> which is odd considering the other 90 elements that occur naturally on Earth exist 'naturally' as solids or gasses only. Mercury can exist also as a solid at temperatures well below $\emptyset°$C. Mercury may be heated to combine with oxygen as mercuric oxide, as gas, but then also as a compound. Whether elemental mercury can actually exist as a gas, when sufficiently heated in the absence of oxygen or in a vacuum is unconfirmed.

35. Bromine…the only other element (#35) to exist as a liquid within life's sweet spot seems to behave a bit like mercury, so in <u>three-only</u> possible states.

36. Space. So what then may be observed in, or literally "said for," the <u>absence</u> of all matter? Space itself seems to exist also in <u>three-only</u> states or conditions, be they:
    a. "flat" or fixed and static,
    b. curved positively or "open," or
    c. curved negatively or "closed."

Discussions and comparisons of the possible states or conditions of space may consume volumes but are currently described by…

37. Spatial geometries…insofar as <u>three-only</u> currently exist within the spatial dimensions to which we are immediately subject:
    a. Euclidian (flat), or
    b. Riemannian (positively curved), or
    c. Lobatchewskian (negatively curved).

Note A: Euclidian geometry, essentially by 'default, is what most non-scientists or non-mathematicians 'know' because it, rather than the other two <u>non-Euclidian</u> geometries, was taught to most all of us in school. Euclidian geometry, as is expressed upon a flat, so <u>two-dimensional</u> surface, so demonstrates and demands:
    a. diameters whose relationship to their circumferences (so when rotated to form circles) is always, exactly pi or a ratio of 3.1416, and…
    b. triangles whose interior angles always total twice 90° or 180°, therefore, and…
    c. lines…given any two or more that are drawn and that intersect any original perpendicularly will remain parallel and not intersect.

Note B: Riemannian geometry is expressed in <u>three dimensions</u> so upon or within a space that is understood to be positively curved. Such space may be regarded as extant upon or within the surface of a sphere. Reasoning being assumed here is that our Big Bang originated as a <u>point</u> (Refer to chapter 2, "Points.") and that it must have expanded since and essentially equally in all directions, so has resulted in an essentially spherical shaped universe, one that expresses positive curvature therefore and results in:

a. diameters of circles (so scribed on a <u>three-</u>dimensional spherical surface or orientation) that do not relate exactly to or via pi. More generally, their relationships, depending upon their lengths, so therefore upon their increasing or decreasing circumferences, always relate to <u>less</u> than pi.

b. the interior angles of triangles as drawn on a positively curved surface that always totals in <u>excess</u> of 180°. Specifically, the interior angles of a triangle drawn by a line from an equatorial base (so at 90°) of a sphere to either pole, then rotated 90° back down to the said base (also at 90°) total <u>three</u> times 90° or 270°, for example. Note here also that: $3 \times 3 = 9$, and $3 \times 9 = 27$... just as $3^2 = 9$ and $3^3 = 27$...so the mathematics of "<u>threes</u>" is evident throughout, and...

c. a line in a positively curved space, so then upon a positively curved surface (again, as upon a surface of a sphere), when intersected perpendicularly, so at 90° by any two (or more) other lines, said lines will <u>not</u> be parallel but will ultimately converge to intersect. It is noted here also that any line, from any point within a positively curved space, will in time, so better in distance, ultimately return to its beginning. This compares to a line in Euclidian, so flat space, that will continue in its original direction forever and not return. Einstein's theories and mathematics notably explain and require lines that return to their beginnings. Our space must be positively curved, therefore.

Note C: Lobatchewskian geometry deals with three-dimensional but negative closed space which is much more difficult to visualize, let alone to explain. Suffice it to say, as limited to and within this discourse:

a. diameters of circles drawn in negative space (so then upon "pseudospheres") relate to their circumferences in ratios always in excess of pi.

b. interior angles of triangles must total less than 180°. And...

c. lines drawn perpendicularly to any baseline in negative space diverge, so spread apart, never to theoretically intersect. So then a line drawn from any point in negatively curved space, becomes irrational at best. It expresses a direction, never to return, that becomes conceptually most difficult. Space with negative curvature explained by Lobatchewskian geometry may be subject to examination in *Fizzicks 201*,...not further herein!

Note D: With the assumption of space expanding and being necessarily curved, not flat, then it must be positively so. Unique then to this assumption, that when extended infinitely, even curved space must become essentially flat at some grand extent nonetheless. So...

Note E: Perhaps Mr. Euclid's geometry, as expressed, is (or will be) ultimately correct anyway? Therefore, from any base consideration, so then from any <u>point</u> of original reference, space must demonstrate one of <u>three</u> possible configurations, either:

 a. flat and explained by Euclidian geometry, or
 b. positively curved and explained by Riemannian geometry (and by Einstein), or
 c. negatively curved and explained, however 'obtusely,' by Lobatchewskian geometry.

So then diameters of circles must relate to their circumferences by (three) ratios either:

 a. exactly pi
 b. less than pi, or
 c. greater than pi. And...

Interior angles of triangles must total either (three):

 a. exactly 180°
 b. more than 180°, or
 c. less than 180°. And...

Lines, as drawn or originating perpendicularly to any baseline, must run in either of (<u>three</u>) directions:

 a. Parallel
 b. <u>C</u>onvergent
 c. <u>D</u>ivergent. And...

Lines from any point(s) whatsoever may demonstrate one of <u>three</u> behaviors:

 a. Extend forever in the same direction (Euclidian)

b. Return ultimately to their beginnings (Riemannian), or

c. Go wherever, so 'somewhere else,' and neither have a definable direction(s) nor a direction of return (Lobatchewskian).

Note F: All the above may seem obvious when comparing the three-only geometrics. So then, how possibly don't four or five or more 'options,' or 'behaviors' or 'constants' exist? Answer...they might if: 1. matter had 4 or more natural states. And/or if...2. four or more spatial dimensions, such as a "depth-away" and a "depth-toward" might imply, or that String Theory were to develop as multiple dimensions (of three) within dimensions. Or...3. space had four or more possible 'states' or 'conditions.'

However...

Note G: For the moment, our base assumptions must be that our very existence must be and is explained by a universe that has three 'choices,' better three possibilities...that it is either:

a. Flat...so will expand and stabilize into a relatively fixed status or condition...and remain so forever. This is a rather unlikely circumstance given the universal dynamics currently observed. Or...

b. Open...so positively curved currently and is expanding and will continue to do so indefinitely. Subnote here: Our universal rate of expansion has been observed to also be increasing, so our best possibility may be, in effect, infinite expansion. Or...

c. Closed...so will expand with positive curvature to a condition of minimum density, whereupon (and when upon) it will yield to its own gravity and reverse its expansion. Such will require also, then a reversal of curvature of space, so then to the negative...hard to imagine. Nonetheless, under this <u>closed</u> possibility, our universe might contract then back to a singularity—being its own, secondary <u>point</u> of infinite density and temperature. (As discussed in chapter 20, "The Opposite of Entropy.")

Given then, the <u>three states</u> of matter: 1. solid, 2. liquid and 3. gas(eous), and...

Then the <u>three</u> states or conditions of space: 1. flat, 2. positively curved, or 3. negatively curved...which are explained by the <u>three</u> applicable geometries and which correspond to...

38. <u>Three-only</u> corresponding states of our universe, being: 1. static, 2. open, or 3. closed. Subnote here, and in a perfect transition to the metaphysical world: While in space, and ignoring any gravity or specific acceleration or velocity, space craft are controlled by three options of control: 1. pitch, 2. roll, and 3. yaw... all 'universal' navigational considerations that must respond to height, width and depth...when anywhere!

39. Perhaps other, even <u>many</u> other, scientific 'triplets' exist? Duos or quartets or quintets are rare. May those 'versions' of <u>threes,</u> as noted above, suffice for now...this as opposed to the many nonscientific, so more behavioral, societal, human-constructed items as noted below. These also exist in 'triplets,' some by rules, others by choice(s) perhaps, but most likely <u>not</u> just by coincidence. May the readers decide which are relevant.

1. Our (so Homo sapiens) standard <u>three</u> meals are (or are some combination thereof):
   a. breakfast
   b. lunch, and
   c. dinner.
2. Which meals are consumed with <u>three</u> basic 'tools':
   a. knife (one dimensional per cut)
   b. fork (two dimensional per 'stab'), or
   c. spoon (three dimensional per scoop)
3. Our days and news and pertinent information is delivered and absorbed (or rejected!) in the:
   a. morning
   b. noon/midday, and
   c. evening.
4. Our tastes (sour, bitter, sweet, salty, etc.) are typically determined or discriminated in/via <u>three</u> levels:
   a. too much
   b. just enough = just right, or
   c. too little = not enough
5. Sound preferences are typically <u>three</u> as well:
   a. too loud = intolerable
   b. just right = tolerable = audible, or
   c. too soft = inaudible
6. Our temperature tolerances, better preferences, whether for our coffee, our bedrooms or autos or workplaces (even of our <u>chiles</u> here in New Mexico), are typically <u>three</u>:
   a. too hot (or too 'picante')
   b. tolerable = enjoyable, or 'just right'
   c. too cold (or too bland!)

7. Touch (as in massages, handshakes, and hugs) also display <u>three</u> preferences:
   a. too hard
   b. just right = comfortable, or
   c. too light = insufficient
8. Friendships are typically categorized <u>three</u> ways:
   a. close, cordial
   b. distant, informal, or
   c. nonexistent or undesirable
9. Accordingly, our likes and dislikes in general are typically <u>three</u>:
   a. like
   b. don't care
   c. don't like = dislike

Facebook allows <u>three</u> basic functions:
   a. friend
   b. unfollow, or
   c. unfriend...yes?

10. Our democratic governments function typically in/via <u>three</u> branches:
    a. executive
    b. legislative, and
    c. judicial
11. Even our law, so our legal outcomes included are <u>three</u>:
    a. guilty
    b. not guilty, or
    c. innocent (logically, if not legally)
12. Our economic levels/categories are typically <u>three</u>:
    a. rich or wealthy = high income
    b. middle class = median income, or
    c. poor = low income

13. Our vehicular traffic is regulated by <u>three</u> required behaviors via <u>three</u> specific colored lights also:
    a. red = stop
    b. yellow = caution
    c. green = go
14. When 'green,' one has <u>three</u> choices to 'go':
    a. right, or
    b. left, or
    c. ahead (Note: Going backward is not a common option at a stoplight!)
15. Most any occasion, when expressing choices or approval or not, we reply in <u>three ways</u>:
    a. "yes," engaged and agreeable
    b. "don't care," disengaged or uninvolved, or
    c. "no," opposed and disagreeable
16. We typically (okay, ideally) live in families of <u>three</u>:
    a. father
    b. mother, and
    c. children (or pets, when the kids move away?)
17. For years, even generations, we have been warned or entertained and/or simply associated with:
    a. parents warning: "I'll count to three"
    b. <u>three</u> strikes meaning "You're out!"
    c. "make three wishes"
    d. "good things come in threes"
    e. "It's easy as 1, 2, 3"
    f. three cheers ("hip hip hurray!")
    g. "this is the third and last time"
    h. three bears

i. three little pigs
j. three blind mice
k. three wise men
l. three stooges
m. three musketeers
n. <u>three</u> Billy Goats Gruff…etc.

All in groups of <u>threes</u>…coincidentally. (Yes…or intentionally?)

Note: It is unknown and possibly illogical why Snow White associated with seven (?) dwarfs or why Santa Claus began with eight (only) reindeer. Fortunately, Santa added Rudolf, so nine (3^2)! It is noted also that the Big Bad Wolf was solo, but he did eat a family of <u>three</u>: Mama, Papa, and Baby Bear!

18. Witness the tripod—often more stable and better balanced than a four-legged structure. Note here: The Egyptians took a tripod lesson in the construction of their pyramids and added a fourth side…to imply time? One wonders (reference: Secrets of the Great Pyramid by Peter Tompkins). This true "Wonder of the Ancient World" demonstrates both its geometry and its architecture and location to be remarkably consistent with the four dimensions of space-time. (Refer to chapter 21, "Pi Is Irrational.")

19. Our Judeo-Christian (I am) first family was (and remains) <u>three</u>:
    a. Mary
    b. Joseph, and
    c. Jesus

Note: Some may disagree…and claim that Adam and Eve had to have been 'everyone's' first family? Accordingly, they must have had one boy <u>and</u> one girl (at least). Admittedly, any other possible 'triplicate' (so, one child only) relationships would be illogical. 'Adam and daughter,' or 'Eve and son,' either to advance our numbers is <u>very</u> unlikely. But then… son(s) and daughter(s), so brothers and sisters…it had to have been either way. It <u>had</u> to have been an incestuous start, however consummated! Sorry to have to be so analytical to and for the 'creationists'!

20. Judeo-Christian (primarily Catholic) lives may expect one of three endings:
    a. purgatory
    b. heaven, or
    c. hell…so beware!
21. And then at death, typically one will be either:
    a. cremated
    b. buried (in the ground), or
    c. slotted (in a mortuary)
    (Okay, a fourth option, "burial at <u>sea</u>," might exist but reserved for Osama Bin Laden!)
22. Our God presents himself as one of <u>three</u> entities:
    a. Father
    b. Son, and/or
    c. Holy Ghost
23. Actually, the Judeo-Christians were not the first at all to perceive its divinities as <u>trinities</u>:
    a. The Egyptians worshiped:
        i. Isis
        ii. Osiris, and
        iii. Horus

    b. For the Sumerians were:
        i. Anu
        ii. Enlil, and
        iii. Ea
    c. For the Babylonians were:
        i. Sin
        ii. Shamash, and
        iii. Ishtar
    d. Homeric Greeks recognized:
        i. Zeus
        ii. Athena, and
        iii. Apollo
    e. Hinduism recognized/worshiped:
        i. the Creator, Brahma
        ii. the Preserver, Vishnu, and
        iii. the Destroyer, Shiva

One could continue, but the point is made…mankind seems to relate to its divine beings in triplicates according to 1. sky/heaven, 2. Earth/local, or 3. underworld/hell…not so different than height, width, and depth…is it? And so, again, three wise-men (only?) visited Jerusalem on or about year Ø—did they not?? Four or five or more apparently did not… and one or two would have been inadequate!

Threes and 1/3^{rds}, however expressed and/or involved, are very evident both physically in our science and universe, and metaphysically in our lives. And both must relate…yes?

Obviously, some pairs and options of twos exist and relate as well. Foursomes…not so much and not so often. But 'triplets' and 'threesomes' are much more stable and more interesting. Groups of three's and third options exist most everywhere, both in nature and in life. Two (or less?) are too often too arbitrary, and four (or more) that relate are again rare (and confusing?).

And then, as far as our physical existence is concerned: Options, conditions, architectures, and stata all appear to present themselves as-in underline three-only possibilities.

So then, and at this point, to one more cosmological comparison and consideration we do experience in our worldly existences…as we are universal travelers upon our planet…

Earth, which is not the first planet (Mercury is), nor second (Venus is) nor fourth (Mars is)…but (coincidentally again?) the…

24. Third rock from the sun…is Earth, our home.

Threes are omnipresent and understandable…and essentially a 'rule' in physics. What they do or have done to and for our lives must relate…somehow? Some may argue for coincidences, curiously similar to the one assumption that our creation and beginning, so our Big Bang was simply "a point of infinite density and temperature" that simply "appeared," and "somewhere" and for "no explainable reason." Coincidences simply don't just happen and definitely do not happen over and over again…especially when observing entire universes.

Relationships in groups of threes and/or 1/3^{rds} do so… over and over again. And if still unconvinced, consider the following parameters of our physical, "causemological" existence(s), again:

25. (Again) Matter. Within the three only spatial dimensions to which one is immediately subject, exists matter. It, matter, so all atomic structures that we know (and, of which we exist), be they

elements or isotopes or organic or inorganic compounds, themselves do exist in three only natural states to which one is immediately subject, so as either:

a. solids
b. liquids, or
c. gases.

26. (Again) Water...is not an element but is an inorganic compound of two hydrogens and one oxygen—of note, coincidentally (?) of three-only atomic structures. Water is unique in its pure state to be able to also exist naturally in all three states of matter, be they as:

a. a solid, so as ice at temperatures of Ø°C or below. Or...

b. a liquid...again its 'natural' state that permits life within the Ø° and 100°C "sweet-spot" as above; Or...

c. a gas, so steam as water is 'vaporized' at temperatures in excess of 100°C.

27. Hydrocarbons or metamorphosed organic materials, commonly noted to be "fossil fuels" as are found also in three primary natural states. Allowing for "bitumen" and "tar sands" that may blur the distinctions, fossil fuels do also exist primarily in three matter states:

a. solid, as in coal
b. liquid, as in oil, and/or
c. natural gas

28. (Again) Earth is notably the third, so neither the first nor second, and obviously not the fourth (again, Mars is) rock from the Sun. And it is basically a rock whose substantive and 'rocky'

constituents (the inorganic and nonaqueous surface portion and nongaseous atmospheric parts) are classified as either:

a. sedimentary,
b. igneous, or
c. metamorphic…<u>three</u> primary 'rocky' existences.

29. (By comparison) Both Earth and the Sun exist also as rather normal planet(s) and star(s) do but that are classically "sized" as being either:

a. small
b. medium, or
c. large

And our planet and star are size-medium, in comparison to like and more distant celestial objects. The same may be said for our solar system and galaxy. We are (and they are) very normal, in that sense, but also very exceptional…as will be disused in their <u>metaphysical</u> settings below…small, medium or large, regardless.

And, in the absence of matter, exists…

30. (Again) <u>Space</u>. So then it exhibits the condition that may be observed given the absence of all matter. Space itself is known to exist in <u>three only</u> 'states' or 'conditions' as:

a. flat, fixed, and "static." Or…
b. curved positively or "open" and expanding. Or…
c. curved negatively so "closed" (and seemingly only if contracting?). TBD.

Discussions and comparisons of the <u>three</u> <u>only</u> possible states or conditions of space may consume volumes but are currently described by…

    31. (Again) <u>Spatial geometries</u>…insofar as <u>three only</u> primary appear to exist:
      a. Euclidian = flat. Or…
      b. Riemannian = positively curved. Or…
      c. Lobatchewskian = negatively curved

Note A (again): Mr. Euclid's geometry #1 may well apply <u>in time</u>…perhaps in the time of dimensional absorption? (Refer to chapter 17, "Dimensional Proportionality.") TBD.

In review: the relevance of <u>three</u> states of universal configuration may be expressed as:

      a. flat and explained by Euclidian geometry. Or…
      b. b. positively curved and explained by Riemannian geometry (and by Dr. Einstein). Or…
      c. c. negatively curved and explained by Lobatchewskian geometry, however obtusely and presumably in a state of universal contraction.

So then in any space, diameters of circles must relate to their circumferences by either:

      a. exactly pi
      b. less than pi, or
      c. greater than pi. Given these <u>three</u> options only and…

Interior angles of triangles must total either:

    a. exactly 180°
    b. more than 180°, or
    c. less than 180°. Ditto, and…

Lines originating as perpendicular to any baseline must run either:

    a. parallel
    b. convergently, or
    c. divergently. Which requires that…

Lines originating from any point(s) whatsoever must demonstrate one of <u>three</u> behaviors:

    a. Extend forever in the same direction = Euclidian
    b. Return ultimately to their beginning(s) = Riemannian, or
    c. Go…wherever, so 'somewhere else' and have neither definable direction(s) nor the possibility of return = Lobatchewskian.

Conclusion: All the above, so the <u>three</u>-<u>each</u> explanations, may seem obvious when comparing the <u>three-each</u> corresponding geometries. So then why again don't more possibly four or five (or more?) options or behaviors exist? Answer (again): they might if any one of <u>three</u> (or more?) other variables were to apply, such as if:

    a. Four or more spatial dimensional <u>directions</u> were to exist, such as a "depth away" versus a "depth toward." But they don't. Or…

b. Space itself might express itself in four or more possible 'states' or 'conditions,' such as being curved simultaneously in <u>multiple directions</u>. But it isn't. Or…

c. String Theory was to display and prove multiple dimensions, within and/or above dimensions, which might confirm:

   i. dimension Ø = ours (so Mine)

   ii. dimension -1 = Tim's, and

   iii. dimension +1 = Herb's…which it might. (Refer to chapter 16, "Herb," and chapter 17, "Dimensional Proportionality.") TBD.

32. <u>The universe</u>. Under any current condition or reasonable assumption, however, our very existence, and so the existence of our universe must again be in one of <u>three states</u>, either:

a. <u>Flat</u>. So it will expand in time and stabilize gravitationally into a relatively fixed or unchanging condition, and remain so forever.

b. <u>Open</u>. So it will maintain positive curvature and expand indefinitely.

c. <u>Closed</u>. So instead, our universe might expand with positive curvature to a point of its minimum density and ultimately lose its inertia and contract.

33. All physical (and related mathematical) possibilities (and relationships) appear to present themselves in <u>triplicate(s)</u>, as are expressed in the following:

a. the <u>three</u> spatial dimensions

b. the <u>three</u> primary energies

      i.   electromagnetic

      ii. nuclear (so combining both the weak and the strong). And...

      iii. gravitational

c. the <u>three</u> possible curvatures...that are explained by...

d. the <u>three</u> corresponding geometries...that describe the...

e. the <u>three</u> possible conditions or 'stata' of space...which allow and are subject to...

f. the <u>three</u> exact speed limits of:

      i.   light,

      ii. gravity, and...

      iii. time...which limit(s) is (are)...or appear to be also for time...

g. exactly 2.99792548×10^8 meters...so <u>almost 3</u> $\times 10^8$ meters per second

All of which must and do for sure exist <u>within</u> the parameters of...

34. <u>Time</u>. And to greatly simplify or better simply to categorize its very existence, time is referenced within <u>three-only</u> possibilities:

    a. the past

    b. the present

    c. the future

Which then best categorizes our most personal stata as:

    a. being born, or

    b. living and being alive, or

    c. being dead

And while alive, we exist therefore as:

    a.   infants/children, or
    b.   adults, or
    c.   the elderly/old

So our consciousness and thinking and understanding are either:

    a.   historic and rememberable, or
    b.   current and observable, or
    c.   futuristic and expectant

As expressed by our:

    a.   memory
    b.   essence/behavior, or
    c.   anticipation/prediction

Which in turn agrees with our understandings of things and events and experiences as being:

    a.   done and over
    b.   in process, or…
    c.   undone, and to do or yet to come

…And the similes, analogies, and metaphors are many but always as measured by and within the three temporal possibilities of again:

    a.   the past
    b.   the present, or
    c.   the future

35. We may now, at this moment in time, again enter our virtual spacecraft while operating in 'zero,' or more accurately in <u>balanced</u> gravity, and pilot or move it according to its <u>three</u> orientations, being again:
    a. pitch
    b. roll, and
    c. yaw

And at this juncture, using time as the transition tool (as it may be), the most obvious <u>physical</u> manifestations of <u>threes</u> have been presented. How then <u>threes</u> may be applicable or how <u>threes</u> may apply both to and within the <u>mechanical</u> and <u>physiological</u> and <u>metaphysical</u> follows to conclude with several additional and more esoteric observations, to wit:

36. The mechanical: Consider some simple 'machines':
    a. A thermostat decides (or better, is set) whether it is either:
       i. too cold
       ii. just right, or
       iii. too hot
    b. A cruise control tells us (better, controls our vehicle) if it's going:
       i. too slow
       ii. at speed, or
       iii. too fast

...and there exist many other, similar, mechanical examples.
    c. A stopwatch. It functions properly, except:
       i. when stopped too early, or

  ii. when stopped too late...doesn't work. But properly...

  iii. when stopped 'on the mark' or 'at the right time'...it does.

<u>Three</u> possibilities, nonetheless.

37. Human (Homo sapiens) female pregnancy certainly existed before calendars but has been confirmed to be (normally and/or ideally) expressed and experienced by:

 a. three trimesters of

 b. three months each, for a total of

 c. nine months (3^2 again!)...and coincides quite nicely with both Earth's rotation and its solar location and its orbit... coincidentally??

38. 'Podium' finishes. In any sport or competition with <u>three</u> or more competitors, <u>three</u> finishes only are truly recognized...first, second, or third. 'Medals are awarded in 1. gold, or 2. silver, or 3. bronze. There is no medal (nor any metal) awarded for fourth (or below)...nor are any places for "fourths" on any podiums so provided! Again, it is the '<u>top three</u>' only.

39. Baseball (our/US's 'national' game) involves nines (3^2 players and innings), but the basis is more fundamentally expressed in multiples of 3's:

 a. three strikes, to produce...

 b. three outs, which advances...

 c. said innings (again $3 \times 3 = 9$ events to produce 9 opportunities)

Also:

    d. Bases are separated equally by 90 (9×10) feet, and

    e. Three bases, 1st, 2nd, and 3rd, exist to represent spatial dimension(s), plus then…

    f. Home base is the 4th to represent time. The game continues (adds time) as long as players reach home base.

Note that:

    g. To make anything 'happen' (as in, for a game to be won or lost), one must score a "run" by adding home base to the other three…and is a metaphor for life, perhaps?

And, note also:

    h. A "3 and 2 count" (per batter) is decided by a (3×2) 6th pitch…and if a "strike," so an unsuccessful result (again, for the batter), he is "out," so end of story (for him). But…

    i. If any 6th pitch is a "ball," the batter is "on," and so the 4th ball, the same as the 4th base, adds time to the game.

Game time seems to mimic space-time in a sport that is played upon a diagonal square or a "diamond," which in reality functions more like a compass or clock-face. Second base, directly in line with the home plate and the pitcher, gets one (a "runner") halfway to his "home plate" and so to score (a "run" by rounding all three bases)…it's a sort of a north-south meridian for the game that points to home "plate" (i.e. not a "base").

If we can expand the metaphors and its mathematics and its geometry…of course, baseball is (must be?) our (the US's anyway) national game. It appears to be (so to have been before any "World Cups") the most notable "<u>World</u> Series"… is it not? And again, as organized in 3's or 3^2's.

The significance of Threes and of One-Thirds need be properly understood to best understand our mathematics, our space, our time, our life, our death…and possibly 'our' baseball?!?

Enough said.

Threes and One-Thirds Are Significant…just as the title of this chapter implies.

Chapter 12

Particle Gender?

It can be a bit misleading, but the basic atomic and force particles are first categorized as either being (A) fermions, those that constitute matter and that take up space; or (B) bosons, those that carry forces and do not take up space. So, for starters:

Fermions	Bosons
protons	photons
neutrons	gluons
electrons	W & Z particles
quarks	gravitons (when
	discovered)
neutrinos	(Higgs too)

As said, the list is confusing insofar as electrons, which do carry electricity, are first fermions because they do have mass and do also take up tiny, yet-to-be measured spaces. Neutrinos too must have ultra-tiny masses and may interact minimally, if at all. All fermions must have some demonstrable and/or measurable mass(es) however. Electricity does create electromotive force, therefore.

Bosons are massless. Photons (discussed later) transmit the <u>electromagnetic</u> forces that exist as gamma rays and x-rays (on the short-wavelength side) and infrared (heat) and radio waves (on the long-wavelength side), and of course, as visible light, blue (shorter) though red (longer) in the middle wavelengths...among others. Electromagnetism is transmitted by photons, so to produce charges as regarded 'pluses' and 'minuses.' Photons also keep electrons, attached to and in touch with their atomic nuclei. Photons do <u>a lot</u>! Gluons and W & Z particles carry the two nuclear forces that keep nucleons in proper architecture and alignment and containment, and their internal 'parts,' primarily quarks, where they belong. The Higgs boson is a very new discovery, and although a <u>boson</u> itself, it may explain mass itself...somewhat an oddity in particle physics as currently understood.

Gravitons have been theorized but are not yet discovered, let alone observed, but are assumed to transmit the force of gravitation.

So much for elementary particle physics...in 'crash-course' format.

The Big Bang was our primal, nucleosynthetic event and created all the basic particles we know and perhaps some that we do not. But initially, it did not create any stable matter, as in no atoms, so no elements that we currently recognize. The event itself was so violent and so hot that a "primordial soup" of the unattached primary particles existed, totally disorganized for some (theorized) ±380,000 years thereafter, according to Dr. Weinstein's calculations. This condition prevailed until the temperature dropped and space expanded and cooled, so the super-hot kinetic energies reduced to the condition that the basic nucleons (again, protons and neutrons) and electrons began to organize and attach themselves as atoms. And the first to form was element number 1, hydrogen, the simplest, and still the most plentiful substance in our universe.

This "phase shift" from disorder (so from the primal soup) to atomic formation (elemental atoms) took again some estimated 380,000 years to begin, which is regarded as the start of "recombination." Subsequent cooling and energy redistribution then allowed the hydrogen atoms to fuse and to combine to form element number 2, helium. There is some evidence that in this process of recombination, some of element number 3, lithium, may have been created as well, but very little was apparently evident. There is no current evidence that element number 4, beryllium, was created during recombination at all. All elements number 4 and up appear to have come later via the subsequent processes of stellar nucleosyntheses. TBD.

Hydrogen has always been the simplest and the most abundant substance in existence, both then and now, and is correspondingly the easiest to understand and to explain per its organization and its architecture. One proton (of electrical charge +1) is its atomic nucleus, which is orbited by one electron (of electrical charge -1). Note that all primary elements contain the same, so an equal number of protons and electrons in each elemental atom, which are electrically neutral, therefore . No neutrons (of electrical charge 0) are attached to elemental hydrogen atoms. Element number 2, helium, has four "nucleons" in each atomic nucleus, so two protons and two neutrons are contained therein. Helium atoms are orbited by two electrons, therefore, again to remain electrically neutral.

All elements are numbered according to the number of protons in each of their atomic nuclei. Element number 1, hydrogen, has one; element number 2, helium, has two; element number 3, lithium, has three…and so on. And, as above, each contains the same corresponding number(s) of electrons to remain electrically neutral. The organizations of elemental atomic structures are really very simple and very logical.

So, what has been learned? Within the 'blast furnace' cores of stars, two hydrogens first fuse together to form one helium, but the hydrogens themselves cannot and therefore do not bring along any neutrons. Instead, two hydrogens (so two protons) meet, and one only sheds or releases a 'positron,' which is a positively charged electron (technically an 'antimatter' particle). Such a release causes that proton to transition to become instead a neutron...which neutron attaches to its then attendant proton and the two unite to form deuterium, which is an isotope, specifically a heavy (and unstable) form of hydrogen. No two-proton-only atomic structures exist. And then the process of helium formation proceeds from that point. This formation of helium and subsequently the formation of all heavier elements (those with additional protons) proceed from there.

This first formation (fusion) of the simplest atomic nuclei to produce heavier and heavier elements, up to and including iron (Fe), element number 26, has been going on subsequently in the cores of stars for eons. And then, when stars collapse (in time, as they do...TBD), the gravitational compaction and subsequent explosion of said stars' iron cores (super novae) create the additional sixty-six heavier and naturally occurring elements up to and including element number 92, uranium. All elements heavier than uranium are/have been manmade (e.g. element number 93, neptunium, and element number 94, plutonium). For the record, both neptunium and plutonium exist on Earth as the result of nuclear (bomb) explosions... both of which bombs, for the record, were developed, built, and detonated in New Mexico and then later over Hiroshima and Nagasaki, Japan, in 1940. We Homo sapiens have created what even our stars did not and do not...a subject for future discussion, perhaps? Nonetheless...

Back to that very first meeting of two hydrogen nuclei (so then of any two protons). They cannot and do not simply "hook up." There is no substance, no element, nor isotope known to consist of just two protons. Instead, and again, one of the two protons first converts to a neutron...but why? Which one "chooses" to convert? Again, only one does!

It's really a simple question: Which (of any two) protons converts, and why? I recently put this question to my wife, Sharon, via a rather crude demonstration with two identical tennis balls on our coffee table. I asked again: "Which one decides or chooses to become a neutron?"

She replied: "The woman does."

I responded: "C'mon. This is not a matter of sex or sexual preference!"

Said she: "Can you be so sure? Women are understood to be the best at making such a momentous decision...yes?"

Said I: "Are you implying that protons demonstrate gender?"

She replied: "Like in most any courtship or marriage, of course they do!"

I thought about this, our exchange about 'gender,' for some time. How possibly? So I did some fundamental research and found that the proton that converts must somehow have a bit higher temperature and that its positron must not be as tightly bound...so it must display more energy, therefore. So I stepped back to realize that a converting proton is a bit hotter, a bit looser, and a bit more energetic. And 'oh my god,' the gender analogy fit perfectly! The more 'female' analogy was obvious!! My wife's observation was absolutely logical!!!

A side note and a bit of subatomic observation here: Protons and neutrons are thought to consist of and so in essence would be made of—three quarks each (basic elemental fermions) themselves…as below. Note also: "Up" quarks contain each a charge of +2/3, and "Down" quarks each contain a charge of -1/3. Arrangements of "up" and "down" quarks make up both these nucleons, as below:

+2/3		+2/3	
+2/3	← and/or →	-1/3	← note the change
-1/3		-1/3	here of 3/3^{rds}

A <u>proton</u> of charge +1   or →   A <u>neutron</u> of charge Ø

It's simple to see that one "up" quark, with the one +2/3 charge, must itself convert to one -1/3 (so negative 1/3) charge, or a "down" quark, for a proton to convert. One only subatomic particle, a <u>positron</u>, has been created and released in this process. So again, it's a one-on-one sort of thing. Since I could not cut open the tennis balls and somehow 'install' quarks for my wife, I left the tennis balls whole. But…

One + 2/3 up quark then must actually first <u>behave as</u> the female…yes? Why else would it convert? That one quark must also be hotter, looser, and more energetic, perhaps? This 'subatomic' analogy appears to fit even <u>more</u> exactly!

The analogy at whichever level stands…maybe? Recall that it is again the <u>female</u> that is perhaps "best equipped to make such a momentous decision"…yes? "Particle Gender" may actually be "Quarkian Gender"…possibly?

<u>Some</u>day, this may make some sense? And…

When/if it does, my wife must take the initial credit regardless. It makes sense that it is first a <u>female's</u> choice.

And to explain further…

Notes on Chapter 12, "Particle Gender?"

1. The analogy is presented according to the process of fusion, type Ia. That is analogous to first a 'masculine' proton that 'couples' with a 'feminine' neutron to form deuterium, with the requirement I being a 'heavy' and 'happy' type of 'coupling,' or isotope of hydrogen. It is noted that the number of protons (so the 'male' factors) in any atomic structure do seem to define the primary 'identity' and/or 'purpose' of any element or isotope. So then the number of neutrons (the 'female' factors) do then seem to define their 'behaviors.' In most stable and less reactive elemental structures, the number of neutrons equals the number of protons…but not always. However, it is generally true that equal feminine and masculine factors are most elementarily "harmonious" in more stable atomic structures. Nonetheless…

2. Element number 1, hydrogen, exists as one proton only with no neutron(s). Hydrogen behaves most like a 'single man'…being aggressive, independent, and opportunistic. Element number 2, however, helium, exists quite peacefully with two protons and two neutrons. With both 'genders' so balanced, helium behaves quite balanced and nonaggressively as well, like a solid family unit. It is nonreactive, as are all noble gases, which it is. Note also helium's "quark equivalency" balances, six up and six down, total twelve as noted. Initial fusion type Ia begins with hydrogen and plateaus first as helium.

3. In reference to the periodic chart of elements, line one: so period one representing the first electron ring—provides eighteen positions or slots (as all lines or periods do), but two only are filled in/on this first line: Element one, with atomic weight one, hydrogen occupies slot one, and

helium, element two with atomic weight four, occupies slot eighteen…and there are no other elemental structures noted in between. Curiously, none exist in between at all on line one with atomic weights two or three, except short-lived <u>isotopes</u>. And only three isotopes so exist therein/upon:

    a. Deuterium, or 'heavy' hydrogen, is the product or union of one proton and one neutron, as generally described in this chapter's text above. Deuterium is the 'happy couple' and the only structure with atomic weight two. It fills requirement I, and its "quark-equivalency" balances, three up and three down, for a total of six as noted.

    b. No structures exist with two protons only nor with two neutrons only. So, 'same-sex' partners are coincidentally absent in or on the periodic chart…in fact.

    c. Tritium, atomic weight three, does exist, if briefly only, as the second, <u>very</u> heavy isotope of hydrogen. It is the union of two neutrons with one proton, and not unlike the classic human "threesome," one 'guy' and two 'gals,' is unstable. It is <u>not</u> a formula for long-term relationships, be they either elemental or social! Tritium does not appear to be directly involved in Type Ia fusion accordingly.

    d. Light helium (He_3), the union of two protons and one neutron, exists after a deuterium 'hooks up' with a spare proton. Light helium exists as the only other (other than tritium) isotope with atomic weight three, and fills requirement II of type Ia fusion with five up quarks and four down, so a total of nine. It is <u>not</u> gender balanced, therefore, and quickly fuses with another light helium to form elemental helium (He_4) and releases two extra or 'spare' hydrogens. Helium, acting in effect as the communal home for two deuterium 'couples,' becomes "quark balanced," with six up and six down, a total of twelve. Upon the release of the two extra proton 'guys,' who then seek out neutron/female partners to re-form new deuterium couples. And the process continues with gender preferences apparently intact... hence a form of continued reproduction via a chain reaction, which is not that different from our own, perhaps?

4. A simple overview of type Ia fusion reveals an apparent "decision." First to form via the fusion of two hydrogens is deuterium. It might follow logically then that two deuterium 'couples' would simply unite, as in a very rudimentary 'double date,' to form helium, but they choose not to do so. If they were to merge so simply, fusion would likely hit a sort of 'pause' or 'end point,' potentially requiring extreme temperatures and pressures that might not exist, and the process would not continue as a chain reaction. So instead...

5. The deuterium couples choose to merge and unite instead with yet other free hydrogens, again to form <u>light</u> heliums (He_3). These more 'female-benefiting' threesome structures merge instead into elemental helium (He_4) with, however crafted, the specific intent of releasing two free protons each from each merger. Then again, said protons with their 'single-man' intentions seek out companions to convert again into deuterium's...and the process (elemental reproduction)

continues. Elemental nucleons' gender preferences are complex and forward-thinking…as if for the intended preservation and furthering of their own existences. This again sounds more than vaguely familiar when based upon the observed intentions of *Homo sapiens!*

6.   Nucleons, protons, and neutrons, the sole constituent particles of atomic nuclei, are themselves constructed of quarks. Quarks are described as unique subatomic structures that can actually exist in six types or six forms that correspond to their specific 'spins' and 'colors,' and all with different masses. Germane only to this discourse are the two only primary (of the six) quarks, the two that construct nucleons:

    a.   the "up" quark with an electrical change of +2/3 each and typically demonstrating 'male' behaviors, and

    b.   the "down" quark with an electrical charge of -1/3 each and typically demonstrating more 'female' behaviors.

A proton consists of two up quarks and one down, so $2/3 + 2/3 - 1/3 = 3/3 = +1$, which is its positive one electrical charge. So a proton displays its masculinity.

A neutron consists of one up quark and two down, so $2/3 - 1/3 - 1/3 = 0$, which is its zero electrical charge/neutrality. And the neutron remains more feminine.

7.   At this point, a deeper gender analogy presents itself. How might a proton mimic a sperm cell? And then, how might a neutron function more like an egg? Never mind—the products of protons and neutrons are ever more complicated (so heavier) but standard atomic structures, and this as compared to more complicated yet quite predictably standardized offspring for sperm and egg. But the functions of their <u>internal</u> structures, so

then of quarks and of corresponding chromosomes, <u>do</u> seem to compare.

8. Comparing quarks to chromosomes:
    a. Two types each, up and down quarks, do exist inside every atomic nuclear particle.
    b. Two types each, X and Y chromosomes, do exist inside all germ cells…both Xs and Ys in male/ sperm cells and Xs only in female egg cells.
    c. For a proton to convert to a neutron (i.e. to change gender), one up quark (again, the more masculine) must "choose" to convert to one down quark (again, the more feminine). And the analogy continues.
    d. When two X chromosomes unite, a female offspring results…more analogous to the neutron result.
    e. When one X and one Y chromosome connect, a male offspring results…more like the creation (however poorly understood) of a proton.
    f. A proton's two-up and one-down quark internal architecture bears a logical similarly to the male gamete's X and Y chromosomal combination. So then…
    g. A neutron's one-up and two-down quark combination must bear a more logical similarity to the X and X, so female chromosomal architecture.
    h. Given the gender identities and the absence of any two proton-only and/or two neutron-only structures, 'same-sex' unions do not occur on an atomic level.
    i. When a proton converts to couple with another proton as deuterium, then its only choice is to

become female. Neutrons do not appear to locate free positions nor to reabsorb them to become protons. 'Transgenderism' on an atomic level appears to occur in the male to female direction only.

    j.   Protons do also, under supernoval conditions, combine with <u>electrons</u> to become neutrons…a process of note more subject to the brute force of supernoval explosion than choice. So again, would resultant neutrons be, in a sense, 'transgender' under such a condition? Apologies if the analogy becomes a bit obtuse! It will happen again when considering 'megatons' and 'negatons."

9.   A proton's 'choice' is again unique and internal… subject again to the 'choice' of one up quark giving up its $+2/3$ charge to accept a down-quark's identity with a $-1/3$ charge. Such a 'sex change' releases/creates a net electrical charge of $+1$, which is granted to (?) or taken by (?) and opportunistic positron. Said positron is emitted (?) or escapes (?) immediately, meets an electron, and annihilates instantly into heat and light. This occurs billions of times per millisecond(s) in our Sun and is why…

10. The <u>Sun shines</u>, and we/life exists therefore. Thankfully, gender identity exists, both on an atomic <u>and</u> on the subatomic level as well!

11. A male germ cell containing both X and Y chromosomes, also unique and internal, chooses to merge with a female germ cell with X chromosomes only. How or why either the X or the Y factor select one another exactly is unknown, but might relate somehow to an up quark's

choice. The 'life force,' what is it exactly? Random? Coincidental? Or purposeful?

12. What is unique are the absences of two potential nuclear particles with either of the following:

    a. Three up quarks, so $+2/3 + 2/3 + 2/3 = 6/3 = +2$, so a particle of such a 'double-proton' charge…potentially a "megaton." Such a super-sexual all-male quark threesome is absent, just the same as any three-proton only structure of atomic weight three is absent as well. Being 'too male' then possibly 'too aggressive' is avoided in particle physics.

    b. Three down quarks, so $-1/3 + -1/3 + -1/3 = -3/3 = -1$, would explain a nuclear particle with such a single negative charge…potentially a "negation." Such a super-sexual all-female quark threesome is absent too as is any three-neutron-only structure in particle physics.

Coincidental? Then what of the possible internal structure of an electron? TBD.

All 'male' and/or all 'female' threesomes do not seem to occur in either atomic or subatomic nuclear architecture…which may speak well for 'nuclear' (no pun intended) family units.

Item	Structure	Symbol	Gender	Chromosomal Equivalency	Quarkian Equivalency	Existence/State
A	Up-quark	$1\uparrow$	male	X	$1\uparrow$	primary fermion
B	Down-quark	$1\downarrow$	female	Y	$1\downarrow$	primary fermion
1	Proton	P	male	XY	$2\uparrow+1\downarrow$	prevalent and independent element
2	Neutron	N	female	XX**	$1\uparrow+2\downarrow$	existent and complimentary
3	(megatron)	(Me)	(super man?)	(XXX)**	$(3\uparrow)$	nonexistent ←
4	(negatron)	(Ne)	(super woman?)	(YYY)**	$(3\downarrow)$	nonexistent ↔ (an electron?)(4)
5	Deuterium	H_2	couple	XY_2	$3\uparrow+3\downarrow$	Requirement I = substantive state
6	(P_2)	N/A	(gay?)	(X_2Y_2)	(N/A)	not permitted ←
7	(N_2)	N/A	(lesbian?)	(Y_4)	(N/A)	not permitted ←
8	Light Helium	HE_3	threesome (1)	X_2Y_4	$5\uparrow+4\downarrow$	Requirement II = interim stage
9	Tritium	H_3	threesome (2)	XY_5	$4\uparrow+5\downarrow$	not required, rare
10	Helium	HE_4	commune (3)	X_2Y_6	$6\uparrow+6\downarrow$	Requirement III = noble element

(1) interim threesome required

(2) interim threesome not required

(3) ... as must be all heavier elements?

(4) possible explanation? TBD in *Fizzicks 201*.

Process of Fusion Type Ia: $2P = P + N + position + P = PNP + PNP = PNPN$ (elemental helium) $+2P$

Note Items: $1 + 2 + 5 = 8 + 2 = 10$ (elemental helium)

** Herein exists a contradiction. If neutrons are 'female,' their chromosomal equivalent must be XX ↔ which would imply the XXX megatron (theorized) to appear/react excessively feminine. Something other must surely occur when adding in another (any?) chromosome? Perhaps this is the key to understanding transgenderism? TBD ... and is best to be addressed by geneticists, not physicists!

13. The same may be said for gametes. No 'triple-chromosomal' arrangements, whether same-sex (XXX or YYY) or multiple-sex threesomes (XXY or XYY) seem to exist under normal circumstances. The 'rules' on the chromosomal level seemingly, and in rare circumstances, can apparently be broken. But the 'rules' on the quarkian subatomic level apparently cannot.

So much for 'gender identities' on either the subatomic (so 'quarkian') or the chromosomal levels. Still unexplained are, under normal, controlled (possibly?) circumstances, these two questions:

    a. Which proton, better which up-quark converts and why as fusion beings? And...

    b. Which gamete, better which chromosomal combination is selected and why via conception?

If for other than 'sexual' preference, then what?

It would be a shame if it were all left up to either some set of 'rules' or even to 'chance.' Imagine a world with no choices at all. So does function follow form only...like a totally robotic or mechanistic existence exclusively? Or...

May instead there be <u>romance</u>...somehow? May form be just influenced by function?

After all, I did ask my wife, and none other(s), to marry. And...

She said "yes." And our behaviors have adjusted accordingly...with two great children, a boy and a girl. We seem to have both chosen and to have behaved logically...and 'nuclearly.'

There <u>must</u> <u>be</u> some sense to this all. My quarks are surely being entertained at least! Hopefully, they will respond accordingly when I revert back to hydrogens and carbons and whatever else 'elemental' remains of me as I reenter the exclusively "fizzical" world!!!

Chapter 13

Dr. Sagan's Apple Pie...aka "Size Matters"

Mathematics and physics are inexorably bound together. Obviously, one cannot proceed, let alone exist, without the other. Mathematics in particular has demanded (and continues to do so) that physical structures with explainable properties must and do "exist." Several examples come to mind.

First might be, in a sense, the electron. What a unique structure it must be. As a fermion, by definition (so by physicists), it must exist and does so with a mass of 1/1836 (more exactly 1/1836.1) of that of a proton, which has been assigned a mass, therefore an atomic weight, quite arbitrarily, of one. So again, the mass/atomic weight of an electron is then 1/1836.1 or .000544633 atomic units...mathematically. No one to my knowledge has stacked 1,836.1 electrons on any sort of scale to see they add up exactly to atomic weight 1.000000000, with both numbers being expressed in thousands of billionths. Accordingly, the dimensions of a proton have been, in essence, measured at $\pm 10^{-13}$cm. But the dimensions of an electron, to my knowledge, have not. Logic would have it, however, that anything that "takes up space" (as electrons do) must also have <u>some</u> dimension(s) nonetheless.

I've yet to see or read either what is the exact internal structure of an electron...but it *must* have one as well, and some internal structure or structures are necessary to explain its exact electrical charge of negative one to exist. Quarks, the structures of protons and neutrons, as in three down quarks (charges of -1/3 each) would make sense, but have not been confirmed. What exactly an electron <u>is</u> is not yet easily explained.

Secondly, consider the photon. What is it exactly? It is generally described as a "quantum packet of light," and light itself is the visible form of electromagnetic energy. So a photon is essentially an arbitrary structure, equivalent to a unit of electromagnetic energy. Its mass is considered to be zero, primarily because it "is" and therefore travels 'at' 186,282 miles per second...effortlessly. Had it any preexisting mass, therefore, it would approach, better it would attain an <u>infinite</u> mass when measured at light (so its) speed...which it does not. Except that it also is better affected, or better lensed, by the curvature of space itself, to the end that it is also retained by the extreme gravitation of black holes...both to be discussed later. Photons seem to respond like other, larger particles <u>with</u> mass, but for different reasons, due to their apparent lack of it.

I've yet to see or to read either what is or might be the exact dimension of a photon, which in itself may be impossible since dimensions also become 0 (or just irrelevant) at lightspeed. Mr. Einstein, were he here, might so explain. Division by zero paradox comes to mind!

Conceptually then, electrons and photons may be regarded as, perhaps, "virtual" particles...but they do and must exist. A similar search is underway for the graviton, which much like the photon carries light energy; the graviton must then, and perhaps in similar fashion, carry the force (energy to be exact) of gravity…but this search has seen no success yet. Gravity <u>waves</u> have recently been confirmed over huge distances, but exactly how they got here has not… not exactly.

Both of these particles (and several others…to be discussed) had to have been present at and therefore had to have been the result of our Big Bang, again our universe's creation. It is interesting to note that the existence(s) of these particles (and again, some others to include neutrinos and gluons and W and Z particles, etc.) are currently without question. And their presences can only be observed by collisions and/or interactions with other particles or with their 'anti' particles…this of course with the notable exception(s) of photons and gravitons! It is hard to imagine an anti-photon, for example. An anti-graviton, perhaps? Or collisions by/with either of them? Admittedly, it's a bit confusing. Nonetheless…

Back to Dr. Sagan's example. He postulated that: An apple pie is not virtual. Cut it in half once, then two pieces remain. Then cut them entirely in half again, which results in four pieces, etc. It is theoretically possible to continue cutting anything in half or into any desired fraction(s) thereof and always half remains…like one-third of one-third or one-tenth of one-tenth, which is by default the mechanism of mathematics dealing with the very small…therefore <u>forever</u>. (Note: This is true for doubling (or tripling) as well. There should be no limit upon how big or small a number could, let alone a thing, nor a structure, might be either.) Back to the one-fourth piece of apple pie…and Dr. Sagan's example…

How many times can this one-quarter piece be halved until one atom, say, of carbon remains? Note here: A professor at Stanford once estimated to me that it would require some thirty-two to thirty-five additional 'cuts' to do so. This estimate seems too high to me, but assuming a very capable knife or other cutting device, how many one-half cuts <u>would</u> be required to pare a pie down to one carbon atom? Regardless...

The carbon atom has six protons and six neutrons in its nucleus when in its elemental form (a form that hopefully does exist exclusively in our pie). Let's assume we can halve the carbon nucleus and disregard any electrons to create a structure containing, then, three protons and three neutrons (so the equivalent of a lithium nucleus). At that time/point, we'd pare that half again in half to produce a structure with either one proton and two neutrons (the equivalent of tritium, ^3H) or two protons and one neutron (the equivalent of light, helium, He^3). And then we must 'cheat' just a bit to then cut either one by one-third (so not by one-half) to permit our cut to then isolate one proton or one neutron. Either way...

One can conceptually imagine paring our pie down to either one proton or to one neutron. Let's examine the neutron...why not? Its internal structure, so its architecture, is thought to (or has possibly been <u>proven</u> to?) consist of one up quark (charge $+2/3$) and two down quarks (charges $-1/3$ each). So then comes our ultimate cut, which is <u>halving the neutron</u>. And what remains? Are we dealing with three half-quarks (depending upon their alignments) or perhaps a quark and one-half in each half? If atoms are—and therefore their constituent particles are and so then are their subatomic particles (quarks)—in fact or in theory the smallest structures that can exist, what then? What then if we can (okay, conceptually) "keep on cutting"? (Refer to chapter 14, "String Theory.") What then might be a half-neutron's electrical charge? Its shape? Its construct?

By pure logic then, what is (are) the smallest dimension(s) if one-half is always one-half...or otherwise, the largest, if double is always double? The concept(s) of infinity (so then "infinitely small" or "infinitely large") come to mind. Mr. Planck has computed arbitrary minimum and maximum sizes and distances. But, infinity itself, either way, large or small, is uniquely a mathematical concept. Frankly, it appears to be illogical when "one-half" and "twice" infinity must logically exist...do they not? To be discussed at another time. (But do refer to chapter 16, "Herb," or chapter 17, "Dimensional Proportionality" for comparable consideration[s].)

Might there ever be a way to pare down an apple pie to a photon? Or to a graviton, perhaps? Probably not. "String Theory" (chapter 14) takes us to and then even beyond currently understood dimensions. But at some limit, things, more specifically the mathematics and laws we know, have to (and probably do) change. And such a change might contradict our ability to 'cut' at all.

Conceptually then, two basic parameters, including the speed of light and possibly the speed of time, must change at or within some dimension(s) as well? And gravity presumably maintains its dimension's same lightspeed at any level. (Refer to chapter 17, "Dimensional Proportionality.") Nonetheless, then the 'speed of time' would have to change. It's complicated! Nonetheless...

Size matters...especially when cutting neutrons in half. Linear accelerators exist so to blast them apart. A "particle knife" or "nucleon slicer" await their invention(s) and construction...the sooner the better in this author's opinion.

Chapter 14

String Theory...An Attempt To First Understand Its Intentions Only

To begin: Select from the verbs that follow;

1. answers	3. creates	5. requires	7. proves
2. demands	4. implies	6. follows	8. confirms

And then: Insert any into either:

A. String Theory _____?_____ String mathematics.

or B. String mathematics _____?_____ String Theory.

Fill in the blanks...either way. Most all sentences that result will be true, again, either way.

Basic String Theory, its initial assumption in particular, is totally understandable: No measurement can be "too small," let alone "the smallest." Plank length, 1.16×10^{-33}cm, can apparently be ignored. But then...

Upon examination of or within such tiny dimensions, <u>what</u> is to be found? <u>Are</u> actual, vibrating "string-like" (actually, more coiled "spring-like") structures present?

Then, if present, what exactly <u>is</u> their construction, and of what? What is their <u>substance</u>...if any? And...

If of "virtual" substance, how can their actual existence be demonstrated, let alone be confirmed or ever <u>proven</u>? Virtual is virtual; actual is actual. Said proofs will be difficult.

All the above are currently unanswered questions.

String mathematics, 'or math' is, or better, <u>are</u> difficult. Reportedly five avenues of inquiry exist currently. It is notable that some of our most intelligent mathematicians have, and continue and will yet begin to, spend their careers developing String Theory's mathematical application(s), as well as seeking its outright mathematical solution.

Its solution is both hoped and intended to be the TOE (<u>t</u>heory <u>o</u>f <u>e</u>verything). Such a theory, itself, <u>in</u> <u>theory</u>, is intended to construct all the equations as "bridges" necessary that will connect:

a. particle theory, as currently understood, with…
b. quantum theory, currently complicated and imperfect (so generating primarily predictions or estimations, not hard answers), and yet being remarkably accurate, with…
c. quantum electrodynamics (QED), which effectively marries particles with mass with nonmassive particles that transmit forces in wave theory…as a unique combination of their two very different existences… and then with and therefore to be explained by…
d. String Theory itself, as it must provide the common structures, which must function as the common denominator(s) to bridge between all four, theories a, b, c, and d, into the TOE.

It is quite a complicated process, requiring the existence of very tiny structures in very tiny spaces. Some author-physicists suggest that String Theory is a bit like Alice's (in *Wonderland*) "rabbit hole" as currently is understood... and that its "tea party" is analogous to a forever <u>un</u>solvable inquiry. Others do not, and encourage and seek both its advance(s) and its ultimate solution.

<u>Super</u>string Theory is suggesting <u>super</u>partners, which have in turn demanded <u>super</u>math. One result (only, to date), the discovery of (actually, the mathematical evidence for) the Higgs boson, produced recently by the Large Hadron Collider at CERN, is encouraging. CERN, currently the world's longest, fastest, most powerful particle accelerator, so then also the world's most sophisticated and, to date, most expensive machine ever built, has accelerated and then collided countless particles at incredible speeds and with massive energies in search of such '<u>super</u>partners'...but without any additional success, to date, much beyond the Higgs result. Superpartners (acronym, SUSYs), like strings, remain theoretical, therefore.

Back to the basic "string" premise, and as may be restated in other chapters herein, it need be understood that: Given any three sticks or rulers or straight lines, any three can be arranged to intersect perpendicularly, so then each one exactly at 90°, and all relative to one another, at any point in the observable universe...as below:

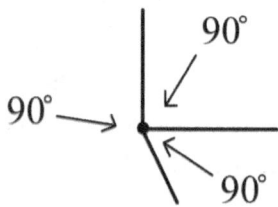

A fourth stick or ruler or line cannot intersect differently and still be perpendicular at that point. Any fourth would connect as a 180° extension of one of the others, which then confirms the three only spatial dimensions being height, width, and depth...as diagrammed above...and again within our observable universe.

So then observable-only spatial dimensions 1, 2, and 3 exist.

Note here that time, dimension number 4 and nonspatial, was added by Dr. Einstein to establish "space-time," which remains the basis of his relativity theories to date. But, the dimensions added in String Theory to date are all spatial.

The above intersection point, so ubiquitous as a point being able to exist anywhere, is considered to be virtually "dimensionless." The intent of String Theory is to redefine such a point as instead having the same three, but also infinitesimally smaller spatial dimensions within. So very tiny height, width, and depth exist also inside the point, and are evidenced then as dimensions 4, 5, and 6. At this/ such a level, time must be reassigned as dimension 7, so that relativity is maintained, if only on a much deeper and much smaller level. Time must not and likely cannot have any spatial measurement at all. But time must be forever divisible as well. How to measure an "instant' is not known...yet.

The existing string mathematics (all five) again are bewildering at these, to date, theoretical levels 4, 5, and 6. And the bewilderment may be increased geometrically when considering superstring theory, which identifies another set of dimensions, so numbers 7, 8, and 9, potentially inside numbers 4, 5, and 6. Under superstring assumptions, time becomes dimension number 10, therefore. As said, time must be "flexible" and is endlessly divisible as above. So then must be space...if String Theory has any merit. Neither space nor time can have any dimensional nor measurable limits.

Therefore, of three observable spatial dimensions, now nine rather than three, are theorized to "have to" exist. And the math associated with 'proofs' of these dimensions as well as the possible <u>contents</u> of such dimensions inside dimensions inside dimensions requires an intellectual and conceptual 'leap' or two at best. Back to the maths...

They (the maths) have become again so complicated and so complex, so essentially to become theories inside theories themselves—that they may exceed human capabilities. So they have been and are increasingly being assigned to computers...so to artificial intelligence—AI, therefore.

It is a safe assumption, perhaps even a safe 'bet' (given wagers are often considered between mathematicians), that AI, so then nonhuman intelligence, will ultimately prove or potentially disprove any and all String or Superstring Theories. This in itself creates unsettling possibilities.

If special dimensions may be so divided and subdivided, so then must time be also. Who can imagine a specific distance that cannot be shortened? Really? Accordingly and by the same logic, who can either imagine a moment of <u>time</u> that cannot be so lessened?

And this logic must apply also in the opposite direction, therefore into larger dimensions as well. Who can imagine a distance <u>or</u> a time that cannot be lengthened or increased? One would suppose that a 'bigness' theory will present itself shortly: "Interstate Highway Theory"? Or something to that end?!?

Future possibilities are fascinating...and so are the 'dimensions' that will have to be evidenced.

Note here that two string theorists, Messrs. Calabi and Yau, have gone on to describe both the theoretical shapes, sizes, and internal architectures of the hypothetical <u>spaces</u> within dimensions 4, 5, 6, 7, 8, and 9. This actually helps define or even to refine the concept of space-time—as in whatever dimensions <u>do</u> exist. (Refer to chapter 17, "Dimensional Proportionality.")

String Theory (or theories?) indeed "get down to the basics." So then...

Its potential criticisms are also worth some consideration. An analogy...

Recall the "Bene Tleilaxu," which were the makers (and sole providers) of artificial eyes for the residents of *Dune* (in the sci-fi trilogy by Frank Herbert...both written and film classics). Did the artificial eyes observe and transmit realities? Or, did they transmit somehow controlled or adjusted or misrepresented images? Colors or sizes or distances were subject to possible manipulation(s). Were any such misrepresentations so designed or caused or controlled by the Tleilaxu, the makers? Or by the eyes themselves? Or perhaps by both?

The novel implied but never really answered these potentialities, I recall. Nonetheless...

It was implied therein, is noted herein, and has been generally evidenced that <u>perception</u> and <u>reality</u> are too often <u>not</u> the same. And this may <u>be</u> evidenced whether observed through our 'natural' or (even) through futuristic, artificial eyes.

Will String Theory computers be possibly analogous to Bene Tleilaxu eyes?

Back on subject: Do or do not Calabi-Yau spaces actually exist? As computed, is the math "real"? As programmed, have the computers printed out their versions...or realities? (Refer to Brian Green's *The Elegant Universe*, chapter 8, page 208 in particular.) So what <u>exactly</u> is one able to observe... the perception (the math and/or computer versions) or the actualities—which <u>are</u> the <u>realities</u>? Recall such spaces are impossibly too small to ever be directly observable. A contradiction may exist therefore.

Considerations and reconsiderations, much like dimensions inside dimensions, can become very esoteric. Some apologies to readers may be due…as herein offered. But…

Without any possible direct examination of even the actual strings or possible springs at all, it is quite evident that unmeasured dimensions inside measured dimensions must exist. What may or may not be found therein must be assigned to a separate consideration and analysis. Such will be addressed in a subsequent discussion…possibly *Fizzicks 201* again.

Conclusion: Who's to "know" for sure? With all of us as humans, our HI (human intelligence) currently makes all final observations and decisions and/or conclusions (this one included), but for how long to come? When will machines (so will AI) become instead the primary observing and decision-making entity? Driverless automobiles followed closely by pilotless aircraft, both currently in existence, even if not yet in general use(s), are here…here to stay. Robots/robotics are still primitive by *Terminator* standards…but for how long?

Virtual reality was promoted chemically, albeit illegally, by Timothy Leary. Now it is sold by any number of companies via perfectly legal goggles and headsets. Try to figure out the math of a "Transformer." Virtual reality is itself rather an oxymoron, as defined. Something must be either virtual or real. It cannot be both…or can it?

So String Theory (or theories), computerized math, Calabi-Yau spaces, and tiny, vibrating "somethings" may or may not actually exist. Perception or reality? Who's to know for sure?

It would appear we are back where we began, so here's the initial set of verbs again:

1. answers   3. creates   5. requires   7. proves
2. demands   4. implies   6. follows   8. confirms

The 'blanks' would appear to remain…? If ever someone is able to provide tangible evidence and then isolate or even actually demonstrate actual "strings" functioning, then Nobel Prize will not be enough!

Computerized answers and TLEILAXU-like images must suffice for some time to come…perhaps?

Look for a more detailed discussion of String Theory in *Fizzicks 201*. Both are projects 'in process.'

Chapter 15

The Absence Of Anything

- Take away the oceans…and there is no surf.
- Take away the atmosphere…and there is no music.
- Take away the _____…and there may be no light nor gravity.
- Fill in the blank.

Were it possible to remove all oceans and water from Earth, obviously no surf nor water 'waves' could exist. Remove the atmosphere and sound waves would not and could not exist. Take away any earthly medium and any earthly propagation of waves of surf or of sound therein is impossible…obviously.

There are places, recently visited by man, where neither water nor atmospheres exist. 'Outer Space,' being simply enough distance from Earth (and external any vehicle), is one. The Moon is another. Neil Armstrong and his successors have confirmed, "There is no surf nor music on the Moon." But light and gravity certainly remain.

There is, therefore, a possible absence that is much more difficult to confirm and is nearly impossible, to date, to prove. Is, in fact, 'empty' space actually empty or not? Any unseen or undetected medium, if extant, was philosophized only early on…so not really 'theorized' and definitely unproven.

Subsequently, a "luminiferous aether" <u>was</u> theorized. Such an "aether" has been debunked by Albert Einstein and some others, as in *A Stubbornly Persistent Illusion* (Einstein and Hawking). In short, Dr. Einstein in particular found that light waves find their own ways through empty space at a constant speed…so they do not <u>require</u> and therefore are not <u>affected</u> by any "aether." He did so in essence by establishing that photons (so quantum packets of light) existed as their own medium, and so they simply organized as their own waves and always at a fixed velocity, "c." Whether this does or does not prove the absence of anything <u>else</u> in empty space is the purpose of this chapter.

Note: The very same considerations can be made for gravity. Dr. Einstein never did tie his theory of gravity (via a space-time continuum) to Quantum Mechanics (the theory of the very, very small), however, and certainly not to String Theory either, which, then unknown to him, has been developed since his death…perhaps to explain quantum mechanics one day? Gravity, though sharing the same velocity, c, of light, behaves differently. TBD.

Nonetheless, it <u>is</u> curious that they both, light and gravity, appear to 'travel' at exactly the same speed…in a vacuum. But the vacuum requirement does not appear to affect the function of gravity. Their 'speed limits' otherwise do emphasize perhaps the most basic natural law that simply exists… that being the cosmic, maximum speed limit of all things at 300,000,000 (estimated) but then 299,792,458 (<u>exactly</u>) meters per second. Stated in miles, said limit is 186,282 mps <u>exactly</u> for all things, including even information (to discuss), in a vacuum. Of note: Gravity appears to proceed, always and everywhere, at exactly the same lightspeed, in and through any and all space(s) and any and all media…never slowed, and always at c as well.

Why this limit is so strictly enforced is unique to our world and our universe. Other natural laws exist and apply, but this one may be the most evident. Refer to chapter 8, "Lightspeed," for some aberrations of light's but not gravity's behavior.

Again, it is quite apparent that light and gravity flow at the same speed and do so apparently unimpeded either through or within (a separate consideration) the same 'empty' space. Light may again be impeded by atmosphere and other transparent materials, however, while gravity apparently is not. Photons (confirmed) flow as light <u>waves</u> (well-established)...much the same as gravitons (theorized but currently <u>un</u>confirmed) must flow as gravity <u>waves</u> (recently confirmed) do. The above gravity waves were and have been predicted, then were demanded by mathematics, and have been confirmed by both experimentation and (now) by observation. Still, that they exist has been recently demonstrated. Exactly how they get here to Earth has not... yet.

The absence of <u>anything</u> in otherwise 'empty' space remains <u>un</u>proven. So then, the same must be said for the opposite/corollary consideration: That the <u>existence</u> of anything in the same space has not been <u>dis</u>proven either. Nonetheless...

Travel to the dark side of the Moon. There is no <u>sun</u>light, but look up at the stars that <u>shine</u>. Anywhere in the blackness of space, one may look in any direction, and plenty of photons are there to enjoy. It remains remarkable, however, that they have traveled from and through enormous distances and over enormous time, and always <u>exactly</u> at lightspeed... so then to be visible here, now, in this, our time. Such consistency is remarkable and commendable...and perhaps even questionable.

Again, how possibly can light exist over any and all times and always behave the same? What allows or causes it to always keep up its exact same speed? When slowed by any intervening atmosphere, it resumes its speed when unrestrained. Can light ever 'slow down' permanently? Can it simply 'become tired'? Light is slowed down through many substances we know…through our atmosphere, through glass or water, to name a few. But assuming there is no atmosphere nor glass nor water between the Hubble Space Telescope and Andromeda, then c is apparently maintained…exactly. It's just 'empty' space again…unless space is not exactly, so not entirely empty.

Gravity waves have recently been recorded from "billions of light-years" away, traveling at the same speed as light, some 671 million miles per hour. Do the math, and the distance is difficult to comprehend. One never knows what might have been encountered en route. Enough said for now.

How about weightlessness in space? Does this not imply the absence of gravity? Weight (which can change) is a measurement of the immediate force of gravity on a specific amount of mass (which does not change). A ten-pound weight on Earth has a mass of roughly 4,540 grams. On the Moon, its weight only will reduce to $\pm 1/6^{th}$ because the Moon's gravity is that much less. But, its mass remains the same. So then in orbit, say halfway between the Earth and Moon, this same 4,540 mass appears "weightless." How possibly?

Gravity is an extremely weak force, when at a distance. Objects, like a ten-pound weight, of comparably 'tiny' mass, may seem weightless when far enough away from any immediate source(s) of gravity, but don't be fooled. The Moon, though apparently weightless too, does not just 'float around.' It remains in orbit up there, just as it pulls on our oceans (and creates tides) down here. Gravity, however weak at whatever distance, is omnipresent. The Moon remains in Earth's orbit, just as Earth so remains in its orbit around the Sun due to gravitational forces at great distances.

So much then for light and gravity. Both are essentially 'always present.' What then other than them, or perhaps including other forms of them, may also be always present in otherwise empty space? To find out, one might go way out, say beyond our solar system, and capture one cubic meter of it...of empty space. And again, ignoring both our light and our gravity discussed above, what if anything else might also occupy this cubic meter?

Take away the _____, and there may be no light nor gravity. So... Fill in the blank.

Noticeably, since Dr. Einstein's passing, the 'blank' has begun to have been filled in. Were the good doctor alive today, his responses would be most interesting to the existence(s) of any of the following four possible additional occupants of 'empty' space:

1. The CMB—cosmic microwave background. It is the remnant heat, therefore the totally extended electromagnetic energy that was blasted out into space by the Big Bang. It is most easily thought of as the faint remainder of the original energy (or energies) created that was (or were) not converted into the atomic and subatomic particles as the result of our primal nucleosynthesis. Early moments of this event, however described, have been condensed into just fractions of a second. Example: The first defined 'Plank phase' has been calculated to have existed for t (minus) 10^{-43} of that first second. The temperature at that instant has been calculated to have been 10^{32} degrees Kelvin. That's t = 1,000,0

00,000,000,000,000,000,000,000,000,000,000 degrees Kelvin, so hot, hot, hot beyond imagination. That same temperature remnant, now omnipresent all around us and essentially everywhere in our universe, is currently measured at ±2.7 degrees Kelvin. Our universe has shed, dispersed, and/or reassigned an enormous amount of energy therefore. The CMB now exists as very, very old and very, very elongated heat waves, so then microwaves that can and have been measured…everywhere. When our universe continues to expand, as it is currently, this remnant heat energy will continue to reduce, ultimately and theoretically, to 0 degrees Kelvin… and will cease to be measurable. So <u>something</u> will have been lost…that is, 'something' that must be included in '<u>anything</u>' currently. It is present now in our cubic meter, regardless. When it 'goes,' what 'goes'? Those remnant photons only? If we were to store our cubic meter separately, say, in a million or a billion years, might any of the remnant photons remain? Recall they travel at ±671 million miles per hour, so where, outside of our cubic meter, might they <u>go</u>?

2. <u>Vacuum Energy</u>…is difficult to describe. Some have attempted to do so as "tension." As the universe continues to expand (again, as it currently <u>is</u> expanding), one may assume it has either of the following:

  a. Enough surplus energy to expand forever. So we would now exist in an 'open' universal mode. Or…

  b. Just enough energy to expand to a point (point of 'balanced' universal density) that is static and will remain so forever. We would exist in a flat or static universal mode if/when that were to occur. Or...

c. Not enough energy to continue to expand forever. So at some future time, our universal expansion might stop, then reverse itself and begin contracting one day. We would then exist in a 'closed' mode that will collapse in upon itself in time...and end in a colossal 'Big Crunch.'

Note: Fortunately, option/mode a, an 'open' universe, seems to be in existence. But given either way(s), a, b, or c above, gravity is and always will be present, even when/if all stars go 'out' and cease to shine...then there would be no more visible light. But the mutual attraction evident now is and always will be evidenced as "tension." So then must Einstein's "fabric of space" and Brian Green's "fabric of the cosmos" actually exist? Therefore, if there is tension, there is 'something' presumably measurable and certainly in existence in our (or in any) cubic meter of space. Gravitons, if and when found, identified, and described, are certainly suspect.

3. Dark Energy...theorized also to (have to) exist. If our universe is behaving 'properly,' as it must behave gravitationally, then the current visible and measurable assumed accountable total mass of what has been so identified explains only ±5 percent of the real total that is necessary. Really? Yes, mathematically. Dark matter, as yet unidentified, is thought to represent ±25 percent of that real total (as below). So then ±70 percent has been assigned, essentially by default, to dark energy. It is interesting to note that if much larger structures, so then even

larger dimensions ("Herbs" universe, refer to chapter 16) do exist, then so must a much larger and grander and all encompassing 'super' gravity also exist...but effectively in 'reverse' from our perspective. Such 'super' gravity, not our own, would logically present itself as 'anti-gravity.' It would be a force pulling our universe apart...therefore causing it to expand more rapidly. In fact, the speed of our universal expansion <u>has been</u> measured recently to be increasing. Measure this with the to-date theorized but not yet identified gravitons, perhaps? There would logically have to be two kinds: our attractive graviton versus a much more dominant retrograde graviton perhaps? Thus may continued expansion continue and "Herb's" world ultimately be in charge? It/they must all be there, both our 'proper' gravitons and all the other 'retrograde' gravitons too, essentially competing in our cubic meter, regardless.

So then, per number 3, assuming dark energy <u>does</u> represent 70± percent of our universe's necessary total mass, then per $E=mc^2$, the king of contradictions also exists: If energy cannot and does not 'take up' any space, it must nevertheless occupy it or be present in it nonetheless. But, given only photons and gravitons to represent the known and theorized greatest component(s), both have been assigned 0 masses to date. So then must some <u>other</u> form of dark energy, possibly anti-gravity, exist <u>with</u> mass? It's difficult to explain 70 percent of a huge amount of unexplained mass with particles that, to date, have none!

4. <u>Dark Matter</u>. This is <u>the</u> suspected but unseen, undetected, and unknown 25 percent of the total real,

necessary, universal mass. It would seem awkward to look for it in/as "dark stars" or as interstellar accumulations of "lightless" dust or of some form of "invisible" gases...or whatever. Black holes, definitely unseen (unseeable) do have tremendous mass(es), but they can't explain enough of the total missing 25 percent. Any/all of the above, currently identifiable stuff will (so must) <u>be</u> present when we examine our cubic meter. However, Dr. Sean Carroll's lecture, "Dark Matter and Dark Energy," poses one explanation: that of "supersymmetry" that might result in "superpartners." An electron's super partner might be a "selectron," not to be confused with its <u>anti</u>-particle, a positron. Might quarks (currently also existing as fermions) have superpartner "squarks" existing as bosons? Photons (currently existing bosons) would/might be superpartnered with "photinos" somehow existing as fermions...so <u>with</u> mass? All the supersymmetry theory has merit, but no observation, let alone confirmation, to date. Dr. Sagan's possible "universes on the tip of my finger" also has some merit...<u>if</u> universal dimensions are plausible. Such a world of 'mini-universes' existing in conjunction, either parallel to or (most likely) <u>integral</u> <u>with</u> <u>ours</u>, certainly would have, however tiny, accumulative mass(es). What 'else' the missing mass might be, it, like dark energy, must have to be inside our cubic meter, regardless. Looking for it somewhere else makes no sense at all. So one must add another 'something,' perhaps another bit of dark matter, however small, in ours and within every other cubic meter of space for sure. Dark matter may be integral

to all space and all matter itself…for that 'matter.'
No pun intended. At this point…

Enter the Higgs boson…boson? Bosons are "force particles," again like photons and gluons (both identified) and both with 0 assigned masses. W^+ and W^- and Z^0 subatomic particles, theorized to provide the 'weak force,' do exist, but do so with smaller masses. Neutrinos too. But again, the Higgs has been observed with a theoretically enormous mass—certainly not typical for bosons! One hundred to two hundred GeVs (billions of electron volts) per Higgs have been estimated. More explanation is needed.

Are we confused yet?

Perhaps, again? However, other exotic forces must exist surely, and some associated subatomic particles must explain them. "Strings" may be just those almost virtual particles… ultimately? Gravitons and potential retrograde gravitons may also be observed and identified in the not-so-distant future. What will be the contribution of AI? It is unknown to date. Needless to say, a great deal has been learned since Dr. Einstein's death. That his explanation of light waves that do not require an 'aether' to propagate through space does not prove the absence, or better, it does not disprove the existence(s) of anything(s) else. Gravitons and dark matter would be again of primary and immediate suspicion.

But again, gravity and gravitons remain inadequately explained. It can be of no doubt they both must exist, along with possibly dark matter and even 'superpartners' in (otherwise) 'empty' space. And then also dark energy…so, possibly, must retrograde or anti-gravity exist as well? Future inquiries…nonetheless.

Recall the wisdom of Native Americans and Central Americans: Navajo and Kuňa today, and Olmec and Mayans of the past. Via their arts and music and stories and histories/ traditions, perhaps even their mythologies as currently understood and however interpreted, they all essentially agree:

"There is no emptiness."
"Something is always present."

The 'blank' would appear to be in process of being filled…yes? And the absence of anything else has <u>not</u> yet been confirmed.

Chapter 16

Herb...et al.

Having been expertly corrected—or better, 'redirected'—by Dr. Carl Sagan, again, some sixty-plus years ago, I have since then sought out and have read a selection of books on <u>theoretical</u> physics...avoiding in particular those that required advanced mathematics. Several of note come to mind:

1. (a) *The Elegant Universe* and (b) *The Fabric of the Cosmos* by Brian Greene
2. *Cosmos* (plus or minus five additional titles) by Carl Sagan
3. (a) *A Brief History of Time* and (b) *The Grand Design* by Steven Hawking
4. *A Universe from Nothing* by Lawrence Krauss
5. *The First Three Minutes* by Seven Weinberg
6. *The Evolution of Scientific Thought* by Abram D'Abro

Nonetheless, Dr. Sagan's own proposition that: "Perhaps I have a thousand universes on the tip of my finger," remains unrefuted. So being totally theoretical, it requires one to imagine the very—as in very, <u>very</u>—small. Less confusing: Imagine just one "universe on the tip of (anyone's) finger," yours or mine. Assume simultaneity and similitude…and the analogy will hold. For starters, imagine a universe one meter in diameter, again "…on the tip of (your) finger." Recall the second (or third?) *Men in Black* movie that involved a universe a bit less than the size of a ping-pong ball worn on a cat's collar? Regardless…

Imagine further, the proportional size of the inhabitants on an imaginary planet within such a universe. We can refer to them as the "Smalls" accordingly. What then would be the Smalls' speed of light? Theirs would require eons of their time to cross their universe, while our light at its speed could do so in an instant. At three hundred thousand kilometers per second, one meter, Smalls' universe requires one three hundred millionths of a second for our light to pass. Comparing the two, Smalls' lightspeed would be essentially infinitely slow compared to ours and would be also in a sense unmeasurable so not to pollute the mathematics and physics we do utilize. And this would lead to a similar consideration and comparison of the Smalls' speed of time as well. TBD.

So if Smalls' existence were to pass for them much the same as ours does for us, therefore, relatively speaking (thank you Dr. Einstein), their speed of time would have to be incredibly fast. In fact, that universe on the tip of anyone's finger might last for only a fraction of a second or even a day or a month or a year or much less…hard to figure or compare. But…

There would exist, from our viewpoint(s), an incredibly slow, almost unobservable speed of Smalls' light, this as opposed to an incredibly fast, perhaps unmeasurable, quickness of their time. To think it through logically, this inverse ratio of lightspeed to time speed would be necessary for 'them' to experience 'their' world the same as we would 'ours.' Refer to note number 7.

Again, such as in "Dimensional Proportionality" (chapter 17), as any interdimensional speed of light <u>decreases,</u> then the corresponding speed of time must <u>increase,</u> which is surely of some interest? It is then perhaps unique (?), that our light's speed (c) must equate, somehow, to our time's speed (t)... again, somehow.

So to make this observation in reverse, imagine instead a world incredibly large. These inhabitants, the "Bigs," would exist wherein their speed of light would be so incredibly fast as to be, now instead and on the upside, essentially beyond our math and physics...arguably 'infinitely' fast. By the same logic then, Bigs' time would also have to be incredibly slow, therefore. It might be/have to be 'infinitely' slower than ours for the Bigs to enjoy their world, again, as we do ours.

Simply stated, again: "Time's purpose is that it keeps all things and events from happening all at once" (Albert Einstein). Smalls' and Bigs' speed(s) of time(s) must then be very different from ours. Again, this is assuming all observations and references are "at rest." Our time (t) must also and does have a constant speed of its own as well, which must be equal to (our) c, all in "Dimensional Proportionality" again (following chapter)...at least as evident in our universe.

If miniature worlds and civilizations and inhabitants of a universe one meter in diameter are difficult to imagine, one might try to think again of the Bigs, so then we must again think in reverse. Apologies for all the imagination and assumed fictions that follow, but the possibility must exist that, and in analogy...

Our galaxy, the Milky Way, may be (analogous to) a molecule or an atom or presumably a subatomic structure, perhaps a quark within a very large object. Pick your object, but mine would be arbitrarily herein a colossal roast-beef sandwich. First, I'd settle upon the specific analogy that our

Milky Way might be a very small part of the beef. And then, if so…

Imagine the size of the person that might be eating that roast-beef sandwich. I have named him HERB, as in "Him Eating Roast Beef." One may assume further that we here on Earth would be totally unaware of having been produced (by a very large beef cow), slaughtered, cooked, packaged, purchased, consumed, digested, and then excreted accordingly…by Herb!

Imagine further then the size of Herb's world, specifically the size of his home planet, and then the size of his star, his sun, and then the almost unimaginable size of his universe. Let's assume, as we already have for Herb and for the (unnamed) Smalls, that the Bigs', his, and the Smalls' existences can be structured much the same as ours and are organized similarly. So then…

Herb's sun would have to be a structure of totally unimaginable size, dimension, and mass and be totally unobservable by Me/us. But over time, his sun in his time, as will our Sun in our time, will have consumed its nuclear fuel(s), and have collapsed via its gravitation in a supernova event of relatively gargantuan proportion…but, then still to a (still dimensional?) point, and so to a possibly 'measurable' singularity as well. So Bigs'/Herb's event would be their Type II supernova even from their viewpoint, and would have resulted simultaneously in conversion of their sun's matter into our energy…analogous again to a black hole in our universe…again via a new, enormous set of Herb's physics, nonetheless. Or via an even 'larger' analogy…

Accordingly, and per the however lengthy evolution of Herb's universe, it too may have originated with too little energy for it to expand at its level +1 forever. So instead, it may have stopped its expansion and, per its gravitation, reversed its course, and over time (a <u>lot</u> of 'our' time), it may have contracted into a supercolossal (very, <u>very</u> big) 'Big Crunch.' So either Herb's universe or his sun contracted to a point, this being a super (almost beyond imaginable) singularity. And that portion, so converted at Herb's end point, is our beginning—our Big Bang.

How else or where else can one imagine the size of a structure that would or could collapse to create us...as in all of us and <u>all of our universe</u>? And this assumes again that our Big Bang was primal. There is understandably another "Big Crunch" possibility of our own universe to discuss. Nonetheless...

Considering then the possibilities of the Bigs (so Herb), the Normals (so Me/us), and again the Smalls (on one's fingertip), it does essentially conjure the possibility of "levels of existence," of universes within universes within universes. And those may as well be new places to look for our dark matter and dark energy. Note that the Smalls' universe's structures themselves could account precisely for the kind of nonreactive, invisible "substance density" being sought to explain our universe's "dark matter" almost exactly. And...

Dark energy. Gravity is and would have to be, via any logic, "multidimensional." It is doubtful that the Smalls' gravity would have much more than its immediate and very confined effect upon them. But the <u>Bigs'</u> gravity would be omnipresent from our viewpoint and (potentially) overwhelming, if evident, and if operating similarly to ours (as must also be assumed), but then also in reverse (retrograde) to ours. Perhaps the perceived acceleration of the expansion of our universe is more easily explained, therefore. The Bigs' gravity would be experienced by us as anti-gravity, pulling our universe apart...ultimately into a "Big Rip"...for later discussion.

So then <u>our</u> universe, currently of ±13.7 billion of <u>our</u> years, exists simply within a never-ending chain of worlds, perhaps similarly organized, perhaps not exactly, but each proceeding at their time(s) and according to their lightspeeds, and all subject to gravitation.

It is evident that any observer in any universe would be aware only of his/her dimensional, so their atomic structures, as well as his/her celestial, so universal structures as well. We all, Big and Small, would be aware of and be able to explain <u>our</u> worlds according to <u>our</u> parameters. Based upon this notion, Dr. Sagan's refutation of any "atom and solar system" analogy may be correct. Their similar organizations here on our level may be simply coincidental. Also, his reasoning that electromagnetic (only) forces are present at our atomic/particle level, and that gravitational force (only) is present at our celestial/cosmic level, may well be correct as well. However...

Our mathematics and physics, so then the "laws" of our world, of our existence, specifically those dealing with the speeds of <u>our</u> light and of <u>our</u> time, would presumably not apply to those other worlds, the universes inhabited by the Bigs' and Smalls'. Their universes would not be just <u>parallel</u>, so "side by side" with ours, but instead, be <u>integral</u> to and <u>within</u> and/or <u>all around</u> ours. There would be therefore no reason to believe in any limits to space or time, therefore. Must they not both expand or contract to include any and all existences that may be, essentially, in both 'directions,' therefore?

If we were to categorize our level, from our perspective, as level zero 0, then the Bigs' would logically be level +1 and the Smalls' level -1. Zero is an interesting number, however, insofar as it is the only number, when added to itself, that remains the same. I'm not (entirely) sure how this affects any reasoning herein! But do refer to note number 7 again at the end of this chapter.

In summary:

Note 1: Given Herb's universal supernova event, if again the collapse of his sun or of his universe, resulted in us/our Big Bang, then ours, were it ever to occur, and presently it appears that it will not, regardless, if our expansion does slow down and reverse, resulting in a Big Crunch singularity of our own, we will have created another 'small bang' relative to ours but resulting in a level (minus) -1 universe nonetheless. Such colossal, so universal, events must create temperatures and densities unimaginably large or small via conventional (our) mathematics...but we must then have to allow for mathematics that would have to adjust for levels +1 and -1 as well.

Note 2: The Smalls' speed of time is, arguably, perhaps infinitely faster than ours. So then ten million per year of their worlds so created might be offset by say slightly less than ten million that cease to exist for any reason at essentially that same time. And...

Note 3: The space we occupy seems to be more than ample to include many, many of the Smalls' universes, especially of a one-meter diameter(s)! So within our dimensional existence level 0, there may be countless integral levels of -1s popping into and out of existence. This sounds much like "quantum jitters" just the same.

Note 4: Speeds (plural) of light can be theoretically understandable. Ours at 186,282 miles per second times 3,600 seconds equals 670.6 million miles per hour. This times 8,760 hours in a year equals roughly 5.874+ trillion miles, which is properly a "lightyear" celestial measurement. So a star fifty lightyears distant would be nearly three hundred trillion miles away. Such is a parameter one might use to consider future interstellar travel (ours), or interstellar communication from any potential extraterrestrial civilization (theirs). Lightspeed will have to be enhanced...somehow? Otherwise, way too much time is involved for anything, including 'communication,' to pass between potential distant inhabitants of just our universe. And this says nothing similarly about 'information' or 'anything else' to pass between universal levels...if and however possible.

Note 5: The concepts of "dark matter" and of "dark energy" do relate rather logically. If the Smalls' worlds do exist, then they, their objects, and their substances would have, however infinitely small, their own mass(es). They/ all bits of their matter would fit in to the current theory of WIMPS, being/behaving as (collectively) weakly interacting massive particles. They would be unseen (so dark) and unable to interact with our matter, but respond collectively to our gravitational forces nonetheless. Such would again be perfect candidates for "dark matter."

Note 6: Our world's gravity does present itself as attracting or pulling in together all the 'stuff' that we know. At the same time, the Bigs' gravity would be pulling all our same 'stuff' apart. Perhaps our universe has and will behave in massive phases of future expansion accordingly:

1. Phase One: having been our primal nucleosynthesis... discussed in multiple references.
2. Phase Two: having been recombination, so approximately 380 thousand years after the Big Bang, when our universe became transparent and visible, therefore.
3. Phase Three: might be referred to as "reconstruction" as the primary atoms and elements formed, giving rise to...

4. Phase Four: when our primal elements (primarily hydrogen and helium) began to clump and coalesce to form suns and planets and galaxies, just as is going on now, while...

5. A sort of phase Five: is occurring simultaneously as gravity begins to slow the explosive inertia, the kinetic energy of our creation. But at the same time...

6. Phase Six: it seems our universe has expanded to a size so that it, and therefore its energy, temperature, and density, has been diluted enough that...

7. Phase Seven: the net effect of the Bigs' gravity eventually becomes stronger than our own, and then...

8. Phase Eight: Our universe expands faster and faster, to demonstrate...

9. Phases Nine and Ten: and...we will need more time to consider these consequences. Nonetheless, we may ultimately exist as "dark matter" for and in level +1, which is Herb's universe.

Hopefully there is, somewhere, another "Herb-like" character that is interested, perhaps even concerned, about us? Figuring the dimensional proportions again, it may ultimately be easier for us to detect the next dimension up, if extant, rather than attempt to find and/or describe the next one down, let alone one among "...a thousand universes on the tip of one's finger." String Theory is attempting to discover and explain those dimensions much smaller than our own. It has added two sets of three or six new ones, so then smaller ones, accordingly. Does the opposite consideration, a "Giant" or an "Immense" theory, also present itself? Why not...since infinity (or infinities) must operate in both directions, must it (they) not?

Time, specifically our time, will tell…if and when we, however unlikely, are around to so observe. Will we wait for assistance from artificial intelligence, perhaps? We cannot fairly continue to assume we are or that we do represent the only rational dimension extant. It may take intelligence vastly superior to ours to discover, let alone to understand, those 'places' or 'universes' infinitely smaller or infinitely larger than our own.

Note 7: It is of note that the three primary phenomena of our level 0 existence and universe are (1) light, (2) gravity, and (3) time, and that all of which we experience must logically travel at the same maximum speed. Light may be slowed while in or traversing through a medium, again like glass or fiber optic cables or our atmosphere. But time and gravity proceed always at our speed, c. May that be, perhaps, how we, our intelligence in particular, is able to travel as well? So…

If time speed (relative to ours) increases and lightspeed (also relative) decreases in a level -1 universe, so then would relative time speed slow and lightspeed increase in a level +1 universe? Accordingly, one might assume that gravitational speeds would then have to behave the same as lightspeeds so to maintain simultaneity. But…

Again, time speeds would behave wherein they would have to change in the opposite(s) and not possibly equate. Herb's/the Bigs' time would be so much slower relatively, as would the Smalls' be so much faster than their respective speeds of light and gravity. How to explain?

This in itself, given the 'equality' in the assumed speeds of our three observed universal phenomena, would seem to present dimensional contradiction(s) of monumental proportion(s). Unless…time, again representing the nonequivalent(s) in speed, is measured only relatively to anyone's observation? Shift the physics (and so the mathematics) perhaps, but would not time's measurements be absolutely relative to "them" while entirely arbitrary to "us"? So actual viewpoints would determine time speeds. Time has no currently understandable 'units' (like photons and gravitons present) to consider. But, to deny any future ability to otherwise quantify or measure time seems unwise… at any dimension.

It would seem that lightspeeds and <u>relative</u> time speeds must and do relate inversely, therefore. If not, again, any assumptions of 'simultaneity' or 'similarity' become difficult to impossible. And gravitational speed may be the common denominator. (Refer to chapter 17, "Dimensional Proportionality.")

Relative 'adaptions' of time speed must exist somehow. So there <u>is</u>, again, fertile material extant for *Fizzicks 201*.

Refer again to a similar discussion in chapter 17, "Dimensional Proportionality," please. Artificial Intelligence (AI) is well on its way…and may reconsider all the above much better than can we? It promises to be a <u>very</u> interesting next fifty to one hundred Earth years, regardless! Hopefully, the "bad actors" currently extant on our planet will behave long enough for our species or our next <u>sub</u>species to figure exactly what, where, and when we are.

Chapter 17

Dimensional Proportionality

It need be understood that Dimensional Proportionality requires 'quantum leaps,' both up and down, in one's conception of space and time. In particular, it requires re-conception of size. Two primary examples are evident:

> First: What was the size, or more specifically, what was the total mass of the structure that must have collapsed or imploded to have resulted in our creation... our Big Bang? To rely only upon the 'happenstance' that a "point of infinite pressure and temperature" simply appeared or existed some 13.7 billion years ago and 'somewhere' is both scientifically and intellectually insufficient. Why and how it appeared, and specifically then what were its origin and grand architecture are the proper inquiries.
>
> Second: As in the 'halving of the neutron' (recall chapter 13, "Dr. Sagan's Apple Pie"), why has logic been apparently neglected? Again, based upon the undeniable rationale

that any bit of matter, or of any substance or of any structure, <u>or</u> substructure can be cut in half indefinitely and therefore be cut an infinite number of times, indicates the frailty of our assumption that our elemental particles and their substructures, be they "quarks" or "strings" or whatever, exist as the smallest possible. Quarks or strings may <u>be</u> the smallest <u>in</u> <u>our</u> <u>dimension</u> that are currently <u>observable</u> and as currently measurable in our universe, but they would not begin to explain the structures necessary and possible in another "universe on the tip of one's finger." (Reference Dr. Sagan's *Cosmos*)

The preceding chapter introduced "Herb," as the theoretical inhabitant of the "Bigs" world. We, in our world, might well identify ourselves as "Me," being <u>M</u>iddle or <u>M</u>edian <u>E</u>ntities. We then can introduce "Tim" being the <u>T</u>iny, <u>I</u>nfinitesimal <u>M</u>an as the theoretical inhabitant of the "Smalls'" world...and then summarize ourselves and themselves and our respective and integral universes as follows:

Name	**World**	**Dimension**	**Size**
Tim	Smalls	-1	TBD ↓
Me	Ours	0	as exists
Herb	Bigs	+1	TBD ↑

What in essence has been demonstrated above is proportional dimensionality, the opposite of Dimensional Proportionality. Either way, the conceptual expressions of organization above are understandable and logical if we introduce and agree to accept 'similarity' and 'simultaneity.'

That is to say that each proposed dimension's cosmic organization must be essentially the same, and that all three (so then the three only that we might examine herein) exist now and at the same, concurrent time(s).

Note: At this point, to conceive of dimensions +2 above or larger, or -2 below or smaller, or of any others beyond may be beyond current comprehension, however, theoretically, may be very logical and probable, nonetheless.

Also, per the prior chapter, the logic of reverse proportionality of the speeds of light versus the speeds of time appears to align perfectly with Einstein's general relativity. Easiest to conceptualize is that as either lightspeed accelerates or that gravity increases (and Einstein equated the two), then time slows down. Accordingly, it makes perfect sense then, that if Herb's dimension exists, our maximum lightspeed must be seen to have been accelerated infinitely to explain his, and his gravity, so then to have his time speed also reduced proportionately.

As dimensions may expand, beginning at whatever level, lightspeeds must increase and time speeds would have to decrease proportionately. Some contradiction may remain, but the base assumptions must remain as being proportional.

Realistically then, if any event measured in our time were to go to zero, which would be our 'moment,' and our minimum time, our lightspeed, which would be our maximum speed, would have to be present and involved…and this would have no effect at all upon Herb's parameter but would demand the same proportionality. So then at his/Herb's lightspeed, his time must also slow to zero. Before addressing the existences of two or more zero times or zero speeds, one must first refute the concept of anti-time, or reverse-time, when/if ever and however our level 0 or any other potential lightspeed were to be exceeded. Such a concept, a reversal of time, remains the stuff of science fiction in or at any specific dimension…for future discussion.

A film or video in any world or in any dimension might be run backward and so be viewed in reverse, but would have no reality in time.

On the mechanical, so non-biological side, and alleging the possibility of reverse time at any level:

1. It is impossible to consider a collapsed star or a black hole ever reforming itself by somehow first 'pushing' its core back out and then 'pulling' it back in to 'reexperience' its supernova, an explosive event in reverse. For a star to somehow 'unfuse' its core and then somehow reconstitute itself in reverse and do so without some sort of reverse supernova is a physical impossibility.

2. To reverse nucleosynthesis on the atomic scale would be to require heavier elements in the cores of stars to somehow decouple and to simplify. In our star, our Sun, for example, elemental helium (essentially a noble gas) might somehow decouple (again) and somehow divide itself back into two deuterium atoms. Whereupon each deuterium's attendant neutron would somehow 'garner up' enough heat and light in a massively endothermic event to somehow recreate positron and electron pairs, which individually would:

    a. add charge +1 to the neutron (a positron function), and/or

    b. fall into orbit (the electron's function). And all said garnering and re-creation of positrons and electrons would again have to be endothermic and incorporate the process, somehow involving anti-gravity…or another nonsensible and impossible mechanism or process in reverse… that cannot and will not occur.

On the living so upon the biological side, possibly being more understandable:

3. A scrambled egg would have to reform, liquefy, and reinsert itself into a shell that 'un-breaks' itself and enters a refrigerator, to subsequently escape back into a supermarket and then into a delivery truck to arrive at the source hatchery and reinsert itself again into its chicken, which has previously been sold to Tyson and also subsequently been made into a chicken taco, which has already been eaten, digested, and excreted…ad absurdum. Or…

4. A sentient being (of any species) would have to 'grow' younger and younger to then become an infant, then become a fetus and reinsert itself into its mother (who may be long since departed much like Tyson's chicken was eaten!), after which its mother becomes un-pregnant, etc., etc., etc…more nonsense.

Note: None of the above can or will occur regardless if or when time might be argued to be able to run in reverse. So then…

These above sequences of reverse-time events cannot and will not happen because time's passage is directionally forward…always and only.

Reverse time(s) and then, by the necessary association, speed(s) in excess of dimensional lightspeed(s) are not permitted, neither proportionally nor on any proportionate level or in/upon any dimension if space times exist at all. Time's 'arrows,' any/all of them, must point forward only. Again, reverse time is the 'stuff' of science fiction. A possible lack of space or of 'anti-space' is similarly not possible...and is not discussed herein.

Accordingly then, in retrograde, if our lightspeed were able to somehow decelerate to zero, our time would speed up to its maximum <u>but</u> <u>not</u> <u>beyond</u>. Speed limits must exist in all dimensions and in both directions and remain in proportion, but again, never in reverse.

Consider any world at absolute zero and essentially at its lightspeed zero...if such a condition might exist. So then certainly any wavelength of electromagnetism would be nonexistent as well. As impossible a situation as this might be, then imagine instead a universe in which, in time, all stars could exhaust their available nuclear fuels and go out. Some (most) would simply convert to (a) black 'dwarfs' and float around much like planets and moons. And the rest might have exploded as supernovae and become either (b) neutron stars or (c) black holes. So visible light and heat would simply cease to be generated, and in time, said universe would appear to have become dark and 'light-less.' So would all be lost? <u>No</u>, not likely.

Still and under any circumstance, both gravity and time would remain. Recall again, that 'weightlessness,' as in the apparent lack of gravity in space, is an illusion. Gravity may become too slight or too balanced to measure, but again, do not be fooled. Gravity at any distance and from/according to any mass, if present, is always present as well...as long as space-time exists. The same goes for time. 'Timelessness,' or our 'zero time,' is only at the point of and for an event at which it currently becomes unobservable and unmeasurable. At our lightspeed, an event in our time passing may go to zero per our mathematics, but time itself must still exist. Back away from any exhibition of zero time, and there still is time. Time(s) in/upon Tim's universe's existence would appear 'instantaneous' or 'simultaneous' per our measurement, but it would allow normal and incremental measurements (normal to the Smalls) nonetheless.

Time by itself must be a condition of the existence of anything and everything in and at any dimension. Time cannot be absent...by definition. So, zero times would appear to be dimensional and mathematical aberrations of measurement(s) of 'moment(s)' only.

And again, time in reverse at any level would not be permitted. So some 'rules' must exist that take precedence over any theory or our current mathematics. These rules would have to be stratified and segmented, curiously similar to the 'lumpiness' of electron wave theory given Plank's explanation and observation of black-box radiation. So, neither 'in-between' speeds of light nor 'in-between' speeds of time may exist. Dimensionality must remain segmented and proportional. Some "universal" rules must exist...somehow.

Reflecting back to a conceptual transition from Tim's world (of the Smalls), then to Me (in ours), then to Herb's (of the Bigs'), lightspeed(s) would exist as either Tim's or Me's (mine) or Herb's, and as none others in between. Same may be for time...either Tim's (infinitely fast) or Mine/Ours (as is) or Herb's (infinitely slow from our perspective), and none others in between would apply.

Accordingly then, as per proper dimensional mathematics, three observed 'zero times' must exist! It is difficult to imagine the existence(s) of three unequal values of zero? 'New maths' will most certainly be necessary.

The above would properly demonstrate Dimensional Proportionality as intended. So, please permit one more mind experiment: Again, the "meter" measurement was initially computed to have been one ten-millionth of the distance from Earth's equator to either of its polar axis points, this assuming, apparently, Earth to be spherical…which it is not. Its equatorial circumference measures ±24,874 miles, while its polar (so perpendicular) circumference measures ±24,860 miles…±14 miles less. Earth is an approximate ellipsoid.

So dividing by 6.2832 (twice pi), the radii are as follow:

- Earth's center to its equator = ±3958.81 miles
- Earth's center to its poles = ±3956.58 miles

Again, the difference is ±2.227 miles and explains an ellipsoid. Earth is 'fatter' at its middle, so when measured perpendicular to its axis…and is not a sphere. However first measured or first calculated, if a meter (which remains an arbitrary, manmade distance) were just a bit shorter, light's speed in a vacuum (fixed and absolute) would measure exactly at 300,000,000 meters per second. Let's assume it does.

So at My/Our lightspeed of 3.0×10^8 meters per second, perhaps…

Herb's might be $3.0 \times 10^{300,000th}$ (plus) power?!?

And so…

Tim's might be $3.0 \times 10^{-300,000th}$ (negative) power?!?

Such measurements may sound fictional, but…

Why not? Until some future genius mathematician (probably an artificial intelligence then, a separate consideration) computes different speed limits for different dimensions, either number however infinitely large (so Herb's) or so infinitely small (so Tim's) must preserve existing Dimensional Proportionality. Mathematics for either world, therefore for either dimension(s), would have to change, or better, would have to be developed accordingly. Dimensionless points in our universe would exist as enormous spaces in Tim's. Herb's universal collapse or simply one stellar supernova in Herb's universe would create his singularity, but with the resulting expression (so dimension) of/for our Big Bang in our universe. 'Black holes' as seen or experienced from the input (or dark) sides would and must result in 'white expressions' or 'big bangs' from the other (or output) sides.

Nothing published that I've read to date actually prohibits the above conceptuality or possibility or even probability—call it/them what you will—of dimensional existences. Again, both 'doubling' or 'halving' of any 'stuff' or of any quantities we know must logically proceed forever and in both directions. Admittedly, conceptualization of Tim's/Small's world is difficult. Coexistence with Me/Our world might either be multiple and parallel (hard to rationalize), or best, integral (reasonably simplified) universes. However...

Conceptualization and theoretical evidence of and for Herb's and the Bigs' world (so singular) is almost demanded, if the description, explanation, and mechanics of our primal nucleosynthesis, our Big Bang, is correct...and if our dimension's dark energy is to be adequately explained.

Lawrence Krauss' *A Universe from Nothing* again assumes an almost proverbial "point of infinite pressure and temperature" that simply "appeared" some 13.7 billion years ago. Such an assumption essentially requires that it, that 'point,' must have simply 'materialized' from 'nothing'...

which remains an inadequate assumption of and for our creation in this author's opinion.

If Herb's universe or perhaps a large star in it did reverse its (assumed) expansion and then did collapse into its colossal or stellar "Big Crunch," then that almost incomprehensible but at the same time perfectly logical event, again for which we have no current mathematical explanation, would explain our Big Bang—to the proverbial "T"! So then any such transition downward (so smaller transition) between dimensions would create subdimensions, accordingly one level down, which does follow logic…if not exactly per our mathematics and physics as currently understood.

If our universe were expanding ever more slowly (as apparently it is not…discussed below), it may well stop in then reverse direction, thanks again to its increasing entropy versus its omnipresent and then also overwhelming gravity, then reverse its direction, and contract into a colossal black hole…like a singularity (again, into our Big Crunch). The other, expressive side of such a singularity, in essence an 'inverse universal' structure, would accordingly produce, or better, would create another of Tim's/the Smalls' universe(s) therefore.

Such would be the reallocation, perhaps reassignment, of our space, matter/mass, energy, and time for them, just as our Big Bang did precisely the same for us. Our energies, both electromagnetic and nuclear, weak and strong, and our 'piece,' or our 'dimensional allocation' of our (possibly) preexisting gravity (to be discussed), would simply be reallocated into their time and existence, again as expressed via Dimensional Proportionality. The elements and functions of our existence as enjoyed by Me/Us would be proportionately reallocated for them. So the downward transition (either Herb's to Ours, or Ours to Tim's) is more easily conceptualized. However…

To provide for or to understand an upward, so an increasing, dimensional transition is more difficult. If all our worlds are best (or just best thought to be) in total dimensional or multi-universal balance somehow, then one must conceptualize the condition and/or mechanics of how one dimension might somehow 'create' or simply 'merge into' so possibly 'be absorbed by' the next larger dimension up as well.

Here's a try that fits well into our current conundrum of our currently 'missing' 25 percent dark matter and currently unexplainable 70 percent dark energy.

Herb's/Bigs' gravity must be evident and eventually may be subsequently overwhelming to Me/Us. Again, to explain our apparently missing dark energy, and excluding all other explanations, Herb's gravity, logically expressed as anti- or retrograde gravity within our universe, makes perfect sense... as does the observation (fact?) that the speed of our universal expansion is increasing. Taken to the ultimate extreme then, our universe might essentially be 'sucked apart' one day by Herb's gravity and so be 'absorbed' and at least 'evidenced' as a part of Herb's dark matter?

Our level 0 existence today would help explain the proper gravitational behavior only of Herb's +1 universe, just as Tim's level -1 universe(s) would explain the current 25 percent missing dark matter for Me/Us. So any next level up's gravitation would explain any lower level's dark energy (or energies) and would govern the rate(s) of their expansion(s). Similarly, the next level down's existence(s) would explain any level up's dark matter necessary for the proper gravitational behavior(s) of any such just higher levels' given existence(s). Read this paragraph more than once so it makes sense! But...

It makes good sense only if the dimensional 'quantum leaps' can be made, either conceptually (so herein) or mechanically to ever, in <u>fact,</u> explain the missing masses and explain proper gravitational behaviors of integrated universes in general.

Again, the physics we know and experience here/within our level 0, including both special and general relativity, seem to work and would seem to apply. At least there do not appear to be any major contradictions, except for the necessary and also proportional 'levels' of the mathematics. 'Infinity,' however large or however small, must be quantified somehow. How then to measure or to observe, let alone to <u>prove,</u> Dimensional Proportionality is a task ahead…presumably for artificial intelligence, which is (hopefully) <u>un</u>impressed by either concept, large or small, of infinity anyway. Computer 'thinking' must not be so limited as is our own. AI need not be introduced to Mr. Plank, bless his soul! Why must we put limits on any quantity or measurement <u>or dimension</u>? Dr. Sagan referred to this, our behavior, as demonstrating our "chauvinism." It would appear that he was correct!

That we may be prisoners of our own mathematics or that we may need to explain new 'mini' or 'maxi' formulations thereof is fertile material for *Fizzicks 201*…which is in process!!

Last notation: Plank's 'constant' of 1.6×10^{-35} meters indicates a minimum distance…this as being the mathematical, not actual, minimum. To imply that $1.6 \times 10^{-35}/2$ does not exist seems totally irrational. What <u>actually</u> prohibits the measurement of both the 'sizes' of spaces 1.6×10^{-70}, for example?

Reconsideration of both the 'sizes' of spaces and then of the 'speeds' of time can vastly expand the field of theoretical physics. And there can be no harm in such thinking outside of any proverbial 'boxes'…however large or small they might have to be constructed!

Chapter 18

Absolute Zero

"Absolute" anything, as well as anything "absolute," seems to challenge its own credibility.

"Absolutism" of most any kind(s) requires all sorts of "proofs of negatives," which have demonstrated themselves to be "unprovable." This is a problem.

Consider this as above, and the concept then of <u>absolute</u> cold, for example. Absolute cold demands the total absence of heat (discussed below), which brings up two (hard-to-prove) negatives. At absolute zero:

1. <u>nothing</u> can become colder, and
2. <u>no</u> heat and therefore no motion can be present...<u>at all</u>.

Absolute zero (so absolute cold) itself has three rather exact measurements in 'degrees,' which are expressed currently as temperatures by common understanding:

1. negative, so -459.67 degrees on the Fahrenheit scale.
2. negative, so -273.15 degrees on the Centigrade (or Celsius) scale, and also...

Per the accepted scientific 'unit' measurement of:

3. 0, as in zero Kelvin units (so not expressed in 'degrees'). Absolute zero is expressed as 0 on the Kelvin scale. Kelvin units correspond exactly to Centigrade degrees otherwise. Water freezes at 273.15 Kelvin, for example.

These are the absolute minimum temperatures 'allowed'...even though none have been exactly found nor reportedly created (yet) in a laboratory: -459.67°F and/or -273.15°C and 0 Kelvin are defined as absolute. Therefore, the following do not and cannot exist:

1. negative, so -460°F or colder
2. negative, so -274°C or colder
3. any sub-zero Kelvin measurements at all, again which can't and don't exist according to an "absolute zero" assumption at 0 Kelvin.

Again, 'negatives' such as these are difficult, if not impossible, to prove.

Cold is the absence of heat. Heat is not the absence of cold. So, exactly how 'cold' can the actual, better, any actual temperature be?

Such is the nature of the absolute minimum temperature that demands the total absence of any heat.

So what then is heat? Four laws of thermodynamics are involved...to include the following:

1. Law number 1 (the conservation of energy): "Energy (heat) can neither be created nor destroyed." This is not an easy concept to accept, so the better, more

understandable statement would be to state: "New or nonexistent energy cannot be created, nor can an old nor existing energy be destroyed." This helps explain Dr. Einstein's famous $E=mc^2$ equation, for example. The "E" in his equation must refer specifically to heat.

2. Law number 2 explains: "In all energy exchanges, if <u>no</u> (again, as in no <u>new</u>) energy enters or if no (again, <u>existing</u>) energy leaves any system, the <u>potential</u> energy of the (ending) <u>state</u> will always be <u>less</u> than that of the initial <u>state</u>." This explains the concept of <u>entropy</u>, as discussed in "The Opposite of Entropy is <u>Contropy</u>," chapters 20, 20(A), 20(B), and 20(C). This law will require some discussion, as herein referenced...with the key words being 'potential' and 'states.'

3. Law number 3 reads: "As temperature approaches absolute zero (our topic), the entropy of a system approaches a constant minimum." By logic then, <u>at</u> absolute zero, a system, therefore <u>any</u> system, can <u>exhibit no further entropy</u>. Be it then, again, that everything and every physical process must stop at 0 Kelvin...which may or may not prove to be 'absolutely' true!

4. Law number 4 is actually referred to as Law number 0 = the "zeroth law," and states (rather obviously): "If two (again, thermodynamic) systems are each in thermal equilibrium with a third one, then they are in thermal equilibrium with each other." Algebra also tells us this:

    a. If $A = C$, and
    b. if $B = C$, then, given $C = C$,

   c.   A = B…which means all the bodies, A, B, and C, must be at the same temperature, obviously. (So, why then the "zeroth law"? Unknown by this author!)

Not as stated (exactly as above) nor as a specific law, <u>also</u> is the concept and observation that: "Any hot body, without any help (as in, without any added heat) will always become colder (as in, will lose its heat) over time." Ergo: Heat will always reduce to become colder by itself, but cannot not similarly increase by itself to become hotter. Therefore, it is always H → C, never C → H, without some outside assistance. Hot to cold is the only 'direction' permitted to occur unaided, without assistance, and by itself. Heat's 'arrow' of existence (or not) is essentially the same as is the arrow of time. Heat's existence is always hot to cold and time's direction is always present to future. But note that, in the case of heat only, a change of direction must exist as well.

It is proposed, again in "The Opposite of Entropy is <u>Contropy</u>," chapters 20, 20(A), 20(B), and 20(C), that gravity would seem to provide the assistance necessary for the formation of stars…a process referred to as "contropy" herein. Gravity essentially reverses entropy…and creates heat therefore.

That heat energy exists or increases (contropy) or decreases (entropy) for all sorts of reasons can be readily demonstrated. <u>But</u> heat's absolute (again) absence cannot. The creation of our universe, theorized (and one must assume) correctly and is demonstrated, really only mathematically, had to have begun with "a point of infinite (so immeasurable by current mathematics) temperature and density"—density being another related topic…for another time. But, our universe began with an 'infinite' temperature, as theorized. Call it "T max." Fair enough, however…

This initial, seemingly 'infinite,' temperature remained so hot that, for ±380,000 (Earth) years, even atomic structures could not form nor could photons organize themselves as light. Understandably, and since then only, did our universe become visible. So, if it is accepted that total age is ±13.72 or ±13.77 billion years old (best estimates of/for the Big Bang), our universe has been visible for some (lesser) 13.34 or ±13.39 years...to be as exact as possible. Let's settle upon the geophysical, or G-year, date of 13.75×10^9 (that is, billion) years when our universe began and G-year 13.365×10^9 when it was able to organize itself and to be seen.

This visibility conundrum will relate to another "absolute" maximum velocity, being that of light, and then of the problems it poses on the upside as well...to be discussed later and elsewhere.

But, coming back to the topic of heat, where has all this initial 'hotness' gone? Answer: "It has gone everywhere and nowhere at the same time(s)...since it's still here. It is just occupying a much larger space." If it was infinite at the exact moment and point of our beginning, it is now measured at 2.725K, relatively consistently and at any point currently throughout our universe. In 'cooling' from 'infinity,' say 10^nK, now to 2.725K, a great deal of this initial, electromagnetic energy (of which heat must be) has dispersed since. And in doing so, its wavelengths have lengthened past infrared, now essentially into (longer) microwave (very much like radio) wave lengths. These much longer wavelengths are still measurable essentially as 'elongated heat' or 'transformed heat' energy on the Kelvin scale, now measurable at 2.725K.

This number, now a 'later-time' measure of our universe's original burst of radiant energy, is referred to as the "Cosmic Microwave Background" via the acronym CMB...or add the word "Radiation," then the CMBR. This same measurement is again at 2.725K average currently throughout our known universe.

Fortunately for us that exist as organic creations (including all plants and all animals), we are living creatures and exist as living organisms here on Earth due primarily to its ambient heat. And two things actually permit us to do so: One is a star that is our Sun that produces our required heat. And two is our atmosphere that insulates us by retaining that heat. And...

This is much the same as saying: "Our atmosphere insulates us from the cold that would prevail otherwise." Our Sun, in effect, heats our atmosphere so to keep Earth's surface temperatures (and most of its real estate) normally between 0° and 100°C. These are again the same temperatures as 32° and 212°F, which are the freezing point and the boiling point of water as measured at sea level. Therefore, liquid water is understood to be an existential requirement for life as we experience it.

Note here: Atmospheric pressure, which is simply the weight as applied by a one-inch-square column of the air above any point on Earth, causes these freezing and boiling points of temperature to change a bit. With increasing elevation (so then with less weight of the less atmosphere above), the freezing point increases a bit, but the boiling point decreases a lot. Example: When/if mountaineering, ask any climber (or observe personally) if it takes less time to boil but more time to cook noodle soup at Camp IV on Mount Everest than it does much lower down in Kathmandu. Actually, one's soup will hardly cook at all if/when on Everest's summit. Water boils at ±162°F at 29,000 feet above sea level...if anyone might like to try! But, lots of soup is cooked instead in Kathmandu!

Regardless, when/if out in deep space, therefore out beyond our Sun's solar sphere, or solar influence, there is essentially no measurable atmosphere, so there is no currently measurable pressure, and the observed ambient temperature is currently again ±2.725K. Note also that water boils almost instantly at zero pressure, so when in space. So, if using the more familiar Fahrenheit temperature scale and the conversion formula: K = 9/5 (F − 32) + 459.67, then to enjoy, say, a 70°F (nice, warm) day on Earth would require 9/5 (38) = 68.4 + 459.67 = 528.07 units Kelvin of increased temperature—so we know! This, again, is our Sun's job… to warm us. Therefore, the temperature drop (again, from a comfortable 70°F) to, say, well outside the orbit of Pluto, would be 528.07 less 2.725 or 525.35K!

Back to more understandable terms, the corresponding decrease in temperature from Earth (i.e. 70°F) to someplace beyond Pluto's orbit (so in deep space) would be roughly 70 + 460 − 2.7, so roughly negative 527°F, which is a most uncomfortable loss of temperature for sure. The absence of heat in space is not conducive to the existence of our life form!

Enough of the mathematics. It is cold, very cold, in deep space. Now to some logic, and…

A thought experiment: Suppose one could construct a box, say 1 × 1 × 1 meter (or roughly 3 × 3 × 3 foot) cube there in deep space. Make the box of super-strong and super-dense carbon fiber material, impervious (so opaque) to any visible light, and 'hermetically sealed' so that nothing elemental (so no matter nor any earthly atmosphere) gets in or gets out. Seal this box shut there somewhere outside of Pluto's orbit, and one would then have a box of essentially nothing, given that deep space is considered to be a vacuum devoid of any 'matter' or other measurable 'atmosphere' at all. And the temperature inside that box would be 2.725K or roughly (negative) -457°F—again inside one's box of 'nothing,' there, just outside of Pluto.

Then somehow bring that box back onto Earth's surface, <u>slowly</u> enough so to avoid any air friction (as experienced by space shuttles and debris coming in at essentially terminal velocities through Earth's atmosphere), but quickly enough to set it down as is, again on a comfortable 70°F day, say on a beach in California, essentially at sea level. Then…

Look up (currently on Google) the rather confusing definition of 'heat'…again per Google (and as per February 2021):

> The universe is made up of matter and energy. Matter is made up of atoms and molecules (groupings of atoms), and energy causes the atoms and molecules to <u>always be in motion</u>—either bumping into each other or vibrating back and forth. <u>The motion of atoms and molecules creates a form of energy called heat</u> or thermal energy, which is present in all matter. Even in the coldest voids of space, matter still has a very small but still measurable amount of heat-energy." (Emphasis added. Note that as herein defined, the motion of atoms (first) "creates…the heat.")

And…

> Energy can take on many forms and can change from one form to another. Many kinds of energy can be converted into heat energy. Light, electrical, mechanical, nuclear, sound and thermal energy itself can each cause a substance to heat up by

> increasing the speed of its molecules. So,
> put energy into a system and it heats up, take
> energy away and it cools. For example...

These definitions (again per Google) tend to ignore that the 'energy' being discussed <u>is</u>, in fact, heat in the first place. And...

Both of these rather poor definitions of 'conduction' and 'convection' of heat fail to specify much concerning 'radiation.'

We have now a contradiction of sorts. We have a container of nothing (again, from deep space) originally boxed at (negative) -457°F, currently sitting on a beach, at a temperature (outside the box) of +70°F, which is a specific temperature difference of roughly 527°F. We assume that the box and its inside space must heat up in time. So it will, in time, assume its surrounding 70°F temperature. But recall also that it is <u>really</u> empty, so it is filled with nothing, and that (again, currently, per Google):

> The <u>motion</u> of atoms and molecules
> creates a form of energy called heat or
> thermal energy, which is present in all
> matter.

The more technical and scientific answers are more complex and have to do specifically with the wave functions of radiant energy. To begin with, energy waves consist of no matter, so they exhibit no mass(es) anyway. And as stated, our box is empty, so it contains "nothing." A basic question, somewhat a conundrum between logic and science, exists therefore: Can "nothing" become hotter or cool down? Can "nothing" possess or itself convey a temperature? Logic would answer, "No. Nothing is what it is and cannot contain or create heat." Nonetheless, there is a residual temperature (again, 2.725K = the CMB) measurable in space. Ergo, something must be present. A trained physicist may go on to explain the transmission of radiant energy through the "nothingness" of space; but again, what about the nonphysicists who can and do ask still "Then what exactly is heat?" or "Why can't I hold onto a red-hot poker or keep my hand in boiling water?" More complicated answers and more questions are coming, for sure.

There is some thought that deep space has no temperature at all, and that "absolute zero" may already exist…somewhere, far enough 'out'…which really depends upon how far 'out' is out…really? Nonetheless…

Heat energy by itself is very difficult to explain, let alone to understand. So…

Back to our 'deep space box' thought experiment again: Let's suppose further that there is (and, in fact, there well may be) an atom of hydrogen (or two) in our box of (essentially, then) nothing. Let's have a look at that one assumed atom of hydrogen. Which consists of…

One electron orbiting one proton—that's it…again, our universe's simplest, most elegant and most abundant element and atomic structure with mass. So a tiny, recognizable unit of matter in our universe, being that one hydrogen atom, is in our box…as is now assumed.

Note here: Any particle physicist would rightly note that even protons, so any hydrogen atom's nucleus, has 'parts' and is made up of 'quarks,' which also have very tiny (so smaller, obviously) masses. But, we can (possibly?) ignore this detail herein when discussing how heat must 'move' or 'agitate' or cause this one hydrogen atom to behave. Fair enough? A more detailed study of quarks will involve quantum mechanics… for *Fizzicks 201*.

Again, our one hydrogen atom has <u>some</u> measure of mass, however small. Its one proton has been assigned an "atomic weight" of one, and its one electron has been measured to have 1/1836th as much mass, or an (additional) atomic weight of .000544. So our one hydrogen atom 'weighs' (again in atomic units) 1.000544 AUs. This can be converted to an understandable (almost) actual weight on Earth and at sea level, of 1.67×10^{-24} grams. So graphically: One atom of hydrogen in our box actually weighs 1,607,000,000,000,000, 000,000,000[th] of a gram. Hydrogen atoms don't weigh much, but they <u>do</u> weigh something, nonetheless.

So, if heat is (and it <u>is</u>) energy, and its energy is expressed via waves of photons—which themselves have no known mass—then heat waves, by definition, have no mass, as in zero. How then does zero mass move or cause an object of even 1.67×10^{-24} grams itself to move? It <u>does</u> and will move, as it must, so to demonstrate heat. No motion, so absolute zero (again and again) demands no heat, therefore no motion. So add heat, and add or cause motion—this is a necessity. But then another contradiction presents itself as well.

How can or how does a wave with <u>no</u> mass cause an object (our hydrogen atom again) with <u>some</u> mass to move (as in, to vibrate)? Back again to our box again sitting on the beach and warming up in California: It now has one (only) hydrogen atom moving inside. Even the super-dense, super-strong carbon fiber walls of the box allow California's ever-present radiant-energy waves (again, of photons) in/ through to cause the one hydrogen atom, confined inside, to move. Given this (albeit, intentionally very specific) state of affairs, it would seemingly take a very long time for this one hydrogen atom itself to heat up our (otherwise) empty space (of, again, one cubic meter) from an original temperature of -457°F to +70°F…would it not?

And again, as stated before, <u>does</u> empty space actually have a temperature? Or, does it instead display an energy <u>potential</u> only?

This, conceptually, must beg the same questions as previously asked herein: Does an actual emptiness of everything and of anything really exist? Is a cubic meter (inside our box again) of deep space (discarding our one hydrogen atom) really empty? How can a 'wave' of anything (so a wave of heat energy) propagate, so move, or so exist in 'nothing'? Don't all waves demand 'something,' therefore some <u>medium</u> in which to move, let alone in which to exist? (Refer to chapter 15, "The Absence of Anything.") Absolute 'nothing,' like absolute zero, is most difficult to define.

Much of this either will or otherwise has been discussed elsewhere in this writing…intentionally. And this bit of difficult-to-explain (yes?) physics has also intentionally been theorized. Okay. Then, let's allow a new perspective, which is…

Our one atom (again) within our box (again), and wherever (again) it might be (again), and at whatever temperature (again) it might be (again), either -457°F or +70°F (again), still consists of a proton and an electron, both with mass(es), and this atom is constantly vibrating, so it is constantly in motion within itself—this is regardless of its location and of its ambient temperature…yes?

Perhaps not. Our lone hydrogen atom displays three, as in three only <u>types</u> of movement(s), which are all, by definition, three <u>sub-types</u> of energetic motion(s):

1. <u>Rotational</u>…in an analogy to Earth (as if it were our lone proton), then Earth's rotation.
2. <u>Transitional</u>…so as compared to Earth's orbit around the Sun, and…

3. <u>Vibrational</u> (which is suspiciously sounding like 'potential'). Think here of a compressed spring that may be itself rotating and orbiting (as above) or not, but its potential or vibrational energy (so its resultant motion upon its expansion, or noncompression) still exists (when compressed).

Quoting several sources then: "At absolute zero, there is (are) no rotational, nor transitional, nor vibrational (so potential?) energy (or energies) at all in any system." And therefore, given a hydrogen atom is perhaps our universe's simplest and most basic 'system,' it must cease to function and must cease to move at all at absolute zero. This means, and therefore <u>must</u> mean, that the one electron moving around our one proton (again) simply stops—stops moving at all... when at absolute zero.

Really? Can or does this occur?

That is indeed difficult to imagine. How and where does (or would or could) any orbiting electron just <u>stop</u>? Perhaps the rotational and transitional energies of the atomic 'unit' essentially freeze up and stop, but the vibrational energies, which are by their very nature, always potential, do not... yes? And again, hydrogen's one electron cannot and will not simply 'freeze' somewhere within its space...will it? Can it?

And further...further into even more basic particle physics: Any heavier atomic nuclei always consist of protons and neutrons—ideally, and when not in isotopic configuration(s), they prefer to exist with the same number of protons and neutrons. Regardless, (and the same is true for all elements heavier than hydrogen)...

Protons have each a positive charge of +1. And all +1s, when in proximity, repel one another. It's not a problem with hydrogen's one and only proton, so when only 'itself' is involved. But consider element number 2, helium. Elemental helium has two protons…and two neutrons. But neutrons have no electrical charge and may be ignored in this example, therefore. They don't repel one another. But again…

The two positively charged protons in a helium nucleus want to come apart. They <u>do</u> repel each other…just as all protons do within any and all heavier elements. So…

What keeps the protons in atomic nuclei together? Answer: <u>Gluons</u> do. Gluons are even tinier subatomic particles, which not unlike photons, appear to be massless. So, per current particle physics theory, energy particles, like photons that behave in/within and as light, so do gluons behave in and around atomic nuclei to contain the protons, and both have (better, <u>display</u>) zero mass(es). Neither photons nor gluons (yet) display any mass(es).

Gluons then, and by definition, represent another form of energy (only four <u>known</u> forms exist <u>currently</u>), one of which is the strong, nuclear force. So again, the known force-energies are:

1. electromagnetic as displayed with/via photons—of no mass(es),
2. strong nuclear as displayed with/via gluons—also of no mass(es),
3. weak nuclear force as displayed by "W" and "Z" bosons—which, theoretically, <u>do</u> have (different yet!) masses, which need be further discussed via quantum mechanics, therefore not further addressed herein. And lastly…
4. gravity <u>assumed</u> to be displayed with/via 'gravitons'— which have yet to have been either observed or identified, but must be assumed to be massless as well. Gravity may or may not be a true "force" energy.

Since the weak-nuclear force, number 3 above, controls radioactive decay in almost all atomic nuclei and is evidenced by bosons (unique subatomic particles) <u>with</u> mass(es), it, again, the weak-nuclear force, is unique and is not (as above) discussed further here in this chapter. Nor can gravity yet be properly discussed as a force, unless/until if or <u>when</u> gravitons are observed, identified, and described. Till then, gravity seems to have to remain a 'quasi-force' energy and so is primarily theorized via relativity only. Back to gluons...

Gluons, again, the transmitters of force energy number 2 as noted above, if and when they are keeping protons in atomic nuclei actually in place and within said nuclei, said gluons must be in constant motion as well. It's rather naïve to assume they are just 'stationary' or just 'there' and/or that they simply do their job(s) while sitting still or while in place. Obviously not. Obviously, they must be in motion, therefore. So then...

What about <u>absolute</u> zero again? Even if electrons were to stop (as in stop rotating, however they might, so stop moving) around atomic nuclei, surely gluons, also at absolute zero, surely cannot themselves simply stop...can they? And note, the same may be assumed for the "W" and "Z" bosons of the weak nuclear force. Otherwise...

Regardless of any temperature, at or conceivably below (if possible?) absolute zero, protons would then be unprotected (again, if gluons stopped too), and said protons would come apart due to their electric repulsion. But here again is the impossibility, the contradiction, and so the absurd situation...

If, under the above assumptions, unprotected/uncontained protons repelled each other and were to come apart, that would require and in fact that would (and currently does) <u>define motion</u>, which <u>creates heat</u>...when theoretically there is none and cannot be any at (or even conceptually, even below?) absolute zero.

At absolute zero, again, if gluons (were to) stop moving, then atomic nuclei are (would be) free to come apart. But they obviously do not. Note: If they did, then all matter in deep space would reduce into free protons...so to exist as disorganized hydrogen nuclei and a bunch of free neutrons only. What a mess...total disorganization! And exactly where would all the electrons go? Let's leave this to a future discussion (again) of quantum mechanics.

However, this/such disorganization would have to at least <u>begin</u> as temperature(s) <u>approach</u> absolute zero...yes? Electrons in particular, and gluons theoretically, would have to at least begin to slow down...would they not?

Electrical repulsion of and between like charges is an electromagnetic force as well. This must be true, just as electromagnetic attraction (as plus and minus charges demonstrate) and magnetism demonstrate as well. So photons would necessarily slow down as well. <u>Attraction</u> is perhaps the key function of all...so of everything! So then, think also of gravity as a sort of non-electrical, so 'uncharged,' attraction. Better yet...

Is gravity also affected at absolute zero? Can or do gravitons (when identified) actually freeze up and stop in the absence of all heat? <u>Now</u> we're getting very <u>close</u> to, if not actually getting <u>into</u>, quantum mechanics again. Really? Because...

Gravity and heat are known to be instead <u>inversely proportional</u>. As any temperature approaches absolute zero, gravity increases, perhaps itself to become maximum or "absolute" itself, if and when in the absence of all heat (i.e., any radiant energy's absence). Yes...and all sorts of new (so different kinds) of 'rules' and 'conditions' seem to exist and to apply in this, a theoretical 'quantum world.' Gravitons could not and logically would not slow down in the absence of heat.

Enter Schrödinger's equation(s)—which, I assume must apply, and which (admittedly) must be understood by/via advanced mathematics that is not able to be presented herein.

So we'll leave this question, better, these questions for another time…but with the last of this chapter's several curious questions:

"Does heat affect gravity?" And if so, then "How does it do so?"

Answer: "Yes." The answer is (apparently?) yes. Gravity varies inversely with temperature. The colder a galaxy (so then a universe too?) becomes, the larger its gravitational force becomes. Then theoretically, such a galaxy (or universe) will eventually contract and implode. This leaves the mechanics of 'how' to future considerations…and so then, by deduction:

When/if a galaxy or a universe <u>were</u> to implode, there would be and must be a lot of motion of a lot of mass(es), which indicates (?), identifies (?), and demands (?) a lot of <u>heat</u>—but again, by the answer just above, there is none, so there is no heat…at least not initially. Therefore, what initially actually occurs within such a process?

More contractions exist…sorry to say. To be continued. Is absolute zero really credible? Or might our measurements of any and all heat be inadequate? Is "absolute zero" actually "absolute"? When does (or when might) heat become evident instead as gravity?

Please refer to the Considerations and Summaries that follow.

Chapter 18A

Considerations And Summary Number One To "Absolute Zero"

Whether it is heat itself that requires motion or that motion produces heat is apparently another "chicken first or egg first" consideration. Nonetheless, heat does move via three distinct mechanisms: conduction, convection, and/or radiation.

Conduction and convection do both require either solid or fluid conductors. Radiation does not. Instead, radiated heat transits or is transmitted via electromagnetic (EM) energy waves, either in infrared or microwave wavelengths, that travel through space, much like light does, and at the same speed. Such radiant transmission is far and away the primary and most prevalent form of universal distribution of heat. Think of our star, the Sun, and of billions upon billions of other stars...all that shine. They all radiate their vast quantities of heat, simultaneously, 24/7/365...just as they are and remain so spread out and so distributed in the cosmos... and have done so for eons.

But they all, or better, it all in essence began at one point and at one time.

Imagine, then, going back to the origination of our universe. It was then, at that moment, as so defined by the "Big Bang," that all matter, or more specifically, all energy burst into an <u>almost</u> instantaneous existence. Hard to imagine? Then consider also that all our recorded time began then as well, which is a subject for an entirely separate consideration.

How and when did such a remarkable event occur?

That event's time would be defined as Gyr 0. Our time now is approximately Gyr 13.75 billion…insofar as the origin of all that we know, so all that is us is understood to have begun precisely at a point in space and in that point of time some ±13,750,000,000 (Earth) years ago.

Subsequently, some 9.25 billion years later, so ±4.5 billion years ago (per current reference), our Sun and Earth coalesced. Our Sun accumulated sufficient mass (i.e., sufficient quantity of matter, so then with sufficient density to create sufficient gravity) to ignite as a star. Fortunately, Earth, being much smaller, did not. Also of note, Jupiter was almost massive enough to ignite…but it did not, again, fortunately for us all! By some estimates, were Jupiter's radius ±20% larger, its core may well have initiated fusion…a discussion for another time!

Again, focus upon that original event, that exact moment at Gyr 0 and exactly at our start time t=0. The original temperature, then T max, had to have been beyond measurement, and so in fact, as if currently by definition, "infinite," again, exactly at that moment. What followed incrementally, first, as measured in seconds or, better, in fractions thereof, then in minutes and then in years, were seven currently agreed-upon and currently identified 'phases' of our very early universe's existence. Said phases, or periods, of our early universe are defined by their approximate times (t's) and then their corresponding estimated temperatures (T's), as so identified.

Phase 1: The <u>Big Bang</u>. The original moment of $t = 0$ and $T = $ infinity (so unmeasurably hot). Not surprisingly, there has yet to be found any known or any quantification of enough heat to explain that original condition. Within heat, again, at that moment, was all energy in combination... worthy of future discussion. (Please refer to a future chapter xx, "Absolute[ly] Hot[ness]," coming in *Fizzicks 201*.)

Phase 2: Inflation...when the singularity that was our starting point displayed its first shape, therefore assumed its first dimension. This occurred within $t = 10^{-35}$ to $t = 10^{-33}$, or within three million, quintillionths (or so) of the first second. The initial (measurable) temperature, then at $t = 10^{-33}$ seconds, has been calculated to have been $T = 10^{28}$ K, or essentially hot beyond comprehension.

Phase 3: Bariogenesis...which would have been the very first formation of primary, massive particles (and of possible anti-particles...a separate discussion), and this process theoretically initiated at $t = 10^{-5}$ seconds and at $T = 10^{13}$K (that's at one hundred trillion units Kelvin)...almost unimaginably hot. Then...

Phase 4: Elemental Pre-Organization...which is understood to have begun at roughly one second and at approximately $T = 10^{10}$K (ten billion Kelvin). Primary nuclear forces, both the weak and the strong, are thought to have separated and to have identified and manifested at this time. Followed by...

Phase 5: Advanced Pre-Organization...basically of hydrogen nuclei (±75% of total) and of helium nuclei (±25% of total), which organized themselves during the first ±100 seconds. Estimated temperature was $T = 10^9$ (that's one billion) K. For comparison, one billion K converts to one billion, 799 million, 999 thousand, nine hundred fifty-four (1,799,999,954) degrees Fahrenheit...to be exact!

Phase 6: Transitional or Intermediate Opaque Era... which has been identified as beginning roughly at $t + 56,000$

years and beyond. Temperatures are thought to have averaged at $T = \pm9000$ K and so (so hot) that did not yet permit photons to properly organize as visible light, nor yet as any longer electromagnetic wavelengths. Then finally…

Phase 7: The "Reorganization"…(which seemingly would more understandably be referred to as the original and visible "Organization") of our early universe. And this has been calculated to have occurred at roughly $t = 380,000$ years. For a comparison, this organization, when previously unorganized photons were finally able to organize as visible light, happened after a period of time (so, subsequent to the Big Bang) roughly equal to the time to date that modern man, aka *Homo sapiens*, has walked upon Earth…350,000+ years!

The manifestation and subsequent expression of EM energies and then the organization of visible light took some serious time…in retrospect. And an enormous amount residual of heat was retained nonetheless…all during these first 380,000+ years.

It remains difficult to imagine even vaguely what that, our very early and for-the-first-time-visible universe, must have looked like…given a current observer's assumed perspective. Few, if any, photons, so essentially, no visible nor organized light had escaped previously. Then, as if magically and as if out of darkness, an immense, boiling cauldron that was our 380,000+-year-old universe must have simply 'appeared.' Notably, it appeared with an average temperature still of $\pm3,000$ K (or $\pm4,937°F$).

Any view, either visual then or conceptual now, so from any perspective would have been and is remarkable.

Nonetheless, this early time (Gyr 3.8×10^5) and early temperature (±3000 K) became the base and original reference points for subsequent and most current universal measurements. Now at this time (Gyr 13.75×10^9), the average residual universal temperature is 2.725 K. This subsequent and actual measurement is referenced now as our Cosmic Microwave Background radiation, with the acronym of CMB (sometimes CMBR). That, 2.725 K average current temperature comparison references and measures the change of/from the ±3000 K computed, so theoretical measurement back at the 'reorganization,' which would have been expressed then via infrared and shorter wavelength EM radiation. It is now measured again at 2.725 K, and is expressed currently and primarily via microwave radiation. The original heat has itself aged and has lengthened and has transformed.

Therefore, in comparison, today, again at 2.725 CMB, our universe, as it has expanded, has redistributed ±3000 K less 2.725 K or ± 2997.75 K worth of heat energy. And in this process, said electromagnetic waves have lengthened generally from the infrared and shorter to the microwave and longer wavelengths. For the purpose of reference, below is the accepted general classification of electromagnetic (again EM) radiation by comparable wavelengths:

Common Descriptions	Wavelengths
1. Gamma rays (original and nuclear energy)	shortest
2. X-rays (penetrating radiation)	shorter
3. Ultraviolet (energetic light)	short
4. Visible (white = all color) light	medium
5. Infrared (invisible heat)	long
6. Microwaves (data and instant heat)	longer
7. Radio waves (electronic communication)	longest

One may logically inquire: "Where has all this lost heat, some 2997.75 K (equal to ±4930°F), gone?" The answer in fact is: "It has gone everywhere and nowhere at and during the same time." In fact, the same total amount of heat has simply 'spread out' to occupy an enormously larger space as our universe has expanded. And in this process of 'spreading out,' it has lost its average energy per cubic (whatever) measurement of the space it currently occupies, and its corresponding wavelengths have become longer, therefore. Infrared and shorter wavelengths have, must, and will continue to lengthen with the passage of time…as long as our universe continues its expansion.

So, one may do some very simple mathematics and divide ± 2997.75 K (lost temperature) by ±13,749,620,000 (years passed since Reorganization) and compute that our universe has lost, on the average, ±.000000218 K (or .000000359°F) temperature per year since its Phase 7 reorganization.

It is then tempting to divide 2.725 K (our remaining heat) by .000000218 K (our average loss per year) to compute roughly 12,614,679, say 12.6 million (more) years, when our universe should cool to absolute zero. Again, this is assuming continued universal expansion and is assuming a consistent rate of cooling. However…

Both assumptions are flawed because:

A. Our rate of universal expansion seems to be accelerating…which would shorten our AZD (absolute zero date). But…

B. As cooling proceeds (as it is and will at any rate[s]), all things, including any said rate(s), slow down as well. Said cooling may proceed, for example, by say 1/2 or by 50 percent per every one million years or so…forever. Therefore…

C. Actually, our universe may get very, very close to but will never actually achieve its AZD.

It would not be difficult (but would be confusing) to compute the estimates of when our universe might reach, say, .000001 K temperature, for example. Again, with continued universal expansion, such a temperature will (theoretically) have to occur on some future date. Unfortunately, it is safe to assume, for *Homo sapiens*, in fact, that no living species will likely exist to make such a measurement. But...

In the process of 'getting there,' so in process of this inevitable transition from <u>any</u> heat at all to <u>almost</u> no heat at all, our current cosmic <u>microwave</u> background (recall our current C<u>M</u>B) will have lengthened further to express itself as a (again, assuming our absence then) cosmic <u>radiowave</u> background. Referring to the seven EM wavelength classifications previously noted. And at some future time, our universe will likely express its residual heat as its C<u>R</u>B)...so via <u>radio</u> waves.

And then what?

      1. how much colder <u>can</u> it get (question 1)?
and 2. how much colder <u>will</u> it get (question 2)?

Question 1: At some point, our universe will approach absolute zero. When it does, or better, <u>as</u> it does, its C<u>M</u>B will have to have reverted to its C<u>R</u>B as above. Any residual heat will then be assumed to be represented by <u>radio</u> waves. Losing the current 2.725 K CMB will take a very long time. Then to perhaps lose its .000001 K C<u>R</u>B (again assumed)...what if any EM wavelength(s) would remain to be 'lost'? The only known wavelength classification above (longer than) number 7 above would be class number 8 (obviously). Class number 7 wavelengths (again, radio wavelengths) are understood to measure from ±1 millimeter to ±100 kilometers...over a very 'broad band,' so to speak. Class number 8 wavelengths measure from ±100 kilometers to 1000+ kilometers and currently both explain and identify as <u>gravity</u>.

So then **Question 2** would be expressed and answered by more complex and theoretical questions themselves: The next longest energetic wavelengths known to exist longer than (EM) radio waves are gravitational. Is this more coincidental than it is scientific? When might or when can electromagnetism express itself as gravity?

Dr. Einstein predicted the existence of very, very elongated gravitational waves some time ago...back in the 1940s, as recalled. And then, as predicted, a set of identified gravitational waves recently crossed the United States from Louisiana to Washington state, therefore in a northwesterly direction relative to Earth's then-current position. Said waves were so detected by two very sensitive Laser Interferometer Gravitational Wave Observatories (acronym: LIGOs). These LIGOs' functional abilities measured the momentary deflections of light beams, as contained within their apparatuses, almost exactly to 10^{-19} of a meter. For comparison, the diameter of a single proton in any and all atomic nuclei is measured on the order of 10^{-15} meters...so ten thousand times larger! And the velocity of these gravitational waves was confirmed to be, as predicted, at lightspeed exactly. The LIGO measurements were extraordinarily accurate...and confirmed that light and gravity travel at the same speed.

The source of these waves, however weakened (reduced in amplitude) they may have become over distance and time, was assumed (confirmed?) to have been the collision and/or merger of two supermassive black holes that had occurred several <u>billion</u> light-years away (and therefore ago)!

Any 'assurance' of this source is obviously theoretical at best, but the same, ultra-longest ever detected gravitational wave set did pass over and around (and assumed through)

Earth not so long ago...specifically back in September of 2015.

So it seems difficult, if not impossible, to imagine, let alone to explain, any electromagnetic wave(s) longer than radio waves. And, simply 'infinite' radiowaves are generated and transmitted by photons...acting as small packets or "quanta" of light (and heat) energy. Longer wavelengths of any such energy at all are again classified as gravitational and would likewise be generated and transmitted instead by gravitons...assumed to be acting as small packets, or "quanta," of said gravitational attraction. The two 'packets,' one being confirmed and identified as photons, and the other being unconfirmed (so hypothetical to date) as gravitons, appear to be very separate entities. Nonetheless...

Is it at all plausible then that EM radio waves, at (or nearly at) or perhaps even below absolute zero would, in effect, cease to exist as such and to somehow convert to either (a) become or (b) behave as gravity? Or simply to (c) give up to gravity? What then, if any, is the relationship that must exist therefore between photons and gravitons?

There must be some connection with reference to the (inverse) observation that gravity increases as temperature decreases. Recent experiments indicate that either gravity seems to displace or to redirect photons, or that photons somehow 'disturb' the passage of (still theoretical) gravitons. However observed, there appears to be an interaction between electromagnetism and gravity...even if only partially understood to date.

Therefore at, or very close to absolute zero, does the 'virtual' absence of heat, therefore the same virtual absence of photons, invite(?) or create(?) or promote(?) or simply allow(?) gravity?

Further discussion is necessary (chapter xx, "Absolute[ly] Hot[ness]," coming in *Fizzicks 201*, will attempt to do so). If the temperature exactly at the moment of the Big Bang was (and currently remains) unknowable, there <u>must be</u> some mathematical measurement or some numerical equivalent that is not simply 'cast off' or that is ignored as being simply 'infinite' (<u>and</u> as being 'unknowable') therefore. It seems logical that...

At some temperature, whether currently immeasurable <u>or not</u>, all four, currently understood energies were either 'contained in,' or were otherwise 'represented by'...<u>heat</u>. Heat, at its (assumed) very hottest <u>upper</u> limit, if existent, appears to be the ultimate and totally expressive and totally radiant inclusive energy form of <u>all</u>. Again...

If the <u>absence</u> of all heat is expressed as absolute zero, then what might be the definition of the <u>presence</u> of all heat? More specifically, what is T max? If light has (and it does have) an absolute maximum velocity, why then can heat <u>not</u> have an absolute and maximum temperature?

As our universe's Big Bang essentially "matured" (again in very minute fractions of its very first second), all the three other currently recognized energies—gravity, the strong nuclear force, and the weak nuclear force—all three simply appeared and became evident. And in whatever order they actually <u>did</u>, they all, all the other three energies, were effectively released and again appeared as the initial heat/ temperature reduced. Therefore...

By logic and however observed, at <u>some</u> maximum (?) temperature, all the other energies must have been previously combined and retained.

What was/is that temperature? Please note that 'infinity' should not be an acceptable answer!

Back to question one: "How much colder <u>can</u> it (our universe) become?" Answer: "It will most likely cool until and if and when gravity becomes dominant." So essentially: "It will cool to <u>nearly</u> absolute zero." And the operative word above becomes 'nearly.' So then to question number two: "How much colder <u>will</u> it get?"

An easy answer: "It will become 2.725 K units colder." But, is this a proper answer?

How much colder—<u>actually</u>, not theoretically? This appears to be a better question…and one can examine the similarities between photons and gravitons to find a partial explanation…perhaps?

It must be more than coincidental that both photos <u>and</u> gravitons (and gluons or mesons, as may be discussed separately as enablers of the nuclear strong force) share several very unique and very distinct characteristics:

1. Both photons <u>and</u> gravitons must be massless, and so
2. Both must also be dimensionless, and in addition,
3. Both display no, as in zero, electrical charge.
4. Both are 'virtual', therefore, and further…
5. Both must (and do) travel exactly at lightspeed.

And again, given the inverse relationship that exists:

6. Gravitation strengthens as temperature reduces.

What then may possibly be deduced?

It/above all begs the question: "Is there a point of possible re-transmission or re-transition of and between electromagnetism and gravitation?" If the Big Bang theories are correct, they (and the two nuclear energies) began united at whatever (yet to be exactly measured) was the initial temperature. Given that, now subsequently, the two nuclear energy forces appear to function separately from any specific residual temperature, might electromagnetism and gravity reunite at absolute zero…or possibly <u>below</u>?

So the simple answer(s) that the universe both 'can' or 'will' lose only its current 2.725 K (only) temperature and/or achieve its AZD (absolute zero date) at some specific, future time are seemingly incomplete and perhaps unreliable…at best.

The grand assumption would have to be that over and above all other considerations, any and all universe(s) (and perhaps even all galaxies?) must at some future time run out of sufficient expansive (therefore, run out of their original heat) energy and therefore cool until and when, even if not exactly, at absolute zero, when gravity becomes overwhelming. In other words…

Cosmic systems, however (and whenever) formed and however originally energetic (read: hot), must ultimately become non-energetic (read: cold) and therefore be without any measurable EMs.

Said systems must then and would then have to reverse their relative disorganization and expansive direction and <u>implode</u>. That is, said systems would then necessarily yield to gravity, then reverse its direction and contract over time again and back into denser and denser organization(s). It is much the same to say that…

Entropy will, at some reduced temperature, reverse itself to become 'contropy' on a massive and cosmic scale. Systems would then contract entirely over time back into gravitational points, then to reform the same super-massive and super-dense, hence <u>super-hot,</u> singularities from which they originated.

Why not? (Refer to chapter 20, "The Opposite of Entropy is <u>Contropy</u>.")

This logical sequence sounds very much like, and would seemingly have to involve, subsequent and so in effect recycling—"Big Bangs." However...

The following are yet to be explained:

A. Which, or better, <u>when</u> was the 'first' one?
B. What then was the first collection of 'whatever' to have imploded?
C. How did it/this all begin?
D. How long has this been going on?
E. Where? (for future discussion)

How possibly can, if it can, <u>heat</u> specifically, not just electromagnetism generally, how can (or does) <u>heat</u> cool or 'lengthen' to either:

1. revert into or
2. become or
3. behave as <u>gravity</u>?

If <u>gravity is</u> the end point of cooling and therefore the end point of entropy, are they (temperature and gravitation) not by definition both 'inversely' <u>and</u> 'universally' related? Is this somehow the same as to say that somehow, photons must revert to behave as gravitons, which must again possibly somehow 'compress' or 'reorient' themselves to become (or to behave as) photons again...is it not? And, there is also 'spin' to consider.

<u>Spin</u> is a unique concept, both scientifically and physically. Its definition: "Spin is the intrinsic angular momentum of any particle or 'quanta,' and is itself quantized

in 1/2 units of 'hbar'." And: "Spin governs the nature of any force." Accordingly, both scientific definitions do result in a physical assumption that particles (and quanta) must have an axis and must rotate either east or west (if viewed face-forward), or clockwise or counterclockwise (if viewed from above or below), etc.

So the relevant nature of any spin(s) might well foretell or describe the direction of any force, therefore. Given that any and all <u>electromagnetic</u> energy is radiant, that is, it seems to flow in an <u>outward direction</u> from its source(s), this as opposed to gravity, which is attractive, therefore is centric only and seems to flow in an <u>inward direction,</u> so then toward its center-of-mass object(s), it must be of some logical, physical significance.

Spin <u>direction</u>, either up or down or right or left or east or west or however spin might logically be described, might well determine the <u>direction</u> of any force so demonstrated? For example: A photon spinning west to east might transmit electromagnetism, while a graviton spinning east to west transmits gravity. And if ever virtual particles are understood, even described to be basically the same kinds of entities, differentiated only by their spin(s)…what then? And, for the record:

Virtual Particles	**Spins**
electrons	½ (observed)
photons	1 (observed or calculated)
gravitons	2 (theorized)

Again, as above, what can be deduced?

It is hard to know, but is also fascinating to surmise! Perhaps photons and gravitons are related by the speed and/or directions of their spin rotations?

Absolute zero may be more a concept than a specific temperature? This question may relate more to the actual spin(s) of a quantum particle going to zero then reversing instead…and spin 'velocities' may vary accordingly?

Does the absence of <u>all</u>, or just the absence of <u>almost all</u>, heat determine our (or any?) universe's future? However answered…

It would appear best to beware of, or simply to avoid the intellectual end-points of <u>anything</u> being "absolute." Perhaps even the spin directions and spin velocities of virtual particles and/or 'quanta' are not absolute?

To be continued in *Fizzicks 201*.

Chapter 18B

Considerations And Summary
Number Two To "Absolute Zero"

Quoted from several sources (this one per Google, February 2020):

> In 2000, the Helsinki University of Technology lab in Finland lowered the temperature of a few atoms even further than researchers [were able to do] in 1995… to the coldest temperature yet reached… to 0.0001 micro degrees Kelvin. <u>But, the atoms continued to vibrate</u>. (Parentheses and emphases added.)

Then, subsequent to 2000 in Finland above…
Quoted recently from *Science News* (also per Google and *Wikipedia*):

> Physicists have now created an atomic gas in the laboratory that nonetheless <u>has negative Kelvin values</u>. These negative absolute temperatures have [display?] several apparently absurd consequences.

> Although the atoms in the gas [then] attract each other and give rise to a <u>negative</u> pressure, the gas does not collapse... (Again, parentheses and emphases added.)

So, two more contradictions become immediately apparent.

<u>First</u>: From the first quote, at a temperature of 1/10,000[th] K (so essentially at absolute zero), atoms do continue to keep functioning. Their electrons (with mass) apparently continue to orbit their nuclei (also with mass) that keep vibrating. That is to say they display their potential energy form in that their 'interior springs' continue to decompress. So...

By any logic, <u>there is motion</u>. And given actual mass(es), so actual matter, there <u>must also be heat,</u> this by any definition. But again, essentially absolute zero demands the essential <u>absence of all heat</u>...how to explain? Atoms should be frozen in time...but they're not. How not possibly?

Our mathematics (specifically of measurements) are perhaps too coarse? Obviously, within 0°K, just as within a point, there must exist smaller-yet measurement(s) of both temperatures, and (just as within) dimensions. Noted...

In String Theory, the existence of smaller and then of smaller yet within smaller yet within (ad infinitum), so then 'curled-up' dimensions must exist. Nothing can be really 'infinitely' small. Infinity over 100 or over 1,000 or over 1,000,000 or over whatever is still a measurement. However, 'theoretical' (so currently unmeasurable) or not, more than the standard three spatial measurements, there being height, width, and depth, must exist. And they must exist essentially no matter how small they become. The dictum: "One can cut anything in half forever...and still half remains" must be unarguably and endlessly true therefore.

So then, consider the same logic with temperature. Absolute zero must apply to perhaps, molecules, but not to atoms? The second quote above again confirms the contradictions of statements noted at the beginning of this chapter. Apparently:

1. <u>Something</u> can be colder (than 0°K), and
2. Some heat is apparently present…always.

So, may we think of temperature like dimension? Perhaps one can 'cut' temperature in half forever, and still half of said temperature remains…also forever?

And so given the above, another pressure contradiction may exist: Given any number of atoms in any space, there must be some pressure, however small. Note: There seems to be a vacuum pressure in deep space, regardless of the existence of <u>anything.</u> (Refer to "The Absence of Anything," chapter 15.) But…

Now beyond (so colder than) absolute zero, even in/with the presence of matter, a <u>negative</u> pressure can apparently exist. How possibly? On second thought, said <u>negative</u> pressure, just above (and per the second quote), sounds suspiciously like <u>vacuum pressure</u>…but is it, or better, how <u>can</u> it then be actually negative? The possible 'directions' of forces may be expressed relative to the possible dimensions involved. TBD.

The implications about temperatures (and about pressures, therefore) here, and additionally about the function of gravitation at this chapter's conclusion, do require further contemplation.

How are negative pressure and gravity related…if at all? Is the possible negative effects of a dimension +1's 'retrograde' gravity so demonstrated and/or explained? (Refer to chapter 16, "Herb.")

Is it perhaps the "quantum mechanics" thing again? Lots of people can <u>explain</u> why it (quantum mechanics) <u>must be</u>, but no one can yet <u>define</u> exactly the way it <u>is</u>!

It must depend upon what the definition of what "is" is...yes? One might argue that this was resolved upon the conclusion of President Clinton's impeachment? Apologies for the pun, as expressed, but the existences of subatomic particles, either identified or virtual, and their relationship is (i.e. remains) a fair domain for future experimentation. And, 'spin' directions and velocities need to be further researched and described.

Chapter 19

Nothings

The play on words about "Nothings" is intentional. Our universe, as currently understood, demands them, so our sciences must embrace them as arbitrarily defined "zero values." Absolute zero (number 1), as discussed in the prior chapters, is the first. There are five other potential "nothings" of note, as below, that warrant examination:

1. electrical charge
2. mass
3. dimension
4. velocity
5. time

These zero-value measures exist, as it appears that they must, to explain the total absences of very specific qualities and/or specific conditions. When and where absent, they determine the starting points for all sorts of events and calculations, processes, and behaviors…again, that are currently understood.

Consider first the net zero electrical charge, noted specifically in atomic and subatomic particles…in particular, in neutrons. Neutrons exist as primary, elemental nuclear 'bits' of matter in all (except for elemental hydrogen) atoms. Neutrons are confirmed to exhibit net zero, that is no electrical charge at all. Why and how are they so?

The constituent, internal parts of neutrons exist as 'quarks.' Exactly three of them apparently combine to construct every neutron. One of any neutron's quarks exists and behaves and is described as an "up" type and carries the fractional electrical charge of positive two-thirds (+2/3). The other two exist as "down" quarks, each with fractional charges of negative one-third (-1/3). So the neutron's electrical mathematics is easily explained:

plus 2/3, minus 1/3, minus 1/3 = 0/3 = 0

Neutrons each exhibit net zero electrical charges, therefore.

This does elicit some questions concerning their behaviors: How and why then do neutrons congregate in atomic nuclei? And more specifically, what initiates or what causes neutrons to do so? Complicated answers result. More neutrons, more than there are protons, or less, less than there are protons, produce atomic isotopes…to be discussed in *Fizzicks 201.*

Recall that of the two (only) primary, massive nuclear particles, which are protons and neutrons, protons seem to be responsible for the basic identities of all elements, while neutrons, again, with zero electrical charge, seem to dictate their behaviors. So, neutron-initiated behavior is of interest… as above and for another discussion.

The other general type, another class of energy particles, exists as 'virtual.' Notable among these are photons (the constituent quanta of heat, light, and other types of electromagnetic energy), gravitons (the assumed constituent quanta of gravity), and gluons (responsible for the strong nuclear force). These primary <u>virtual</u> particles all "exist" also without any electrical charge(s).

Any constituent, so any inner 'parts' of any virtual particles, are specifically unknown to date, and presumably, none exist. So, virtual particles, or 'quanta,' exist (actually, have been defined) as solely their own, unique entities. And in addition to lacking any electrical charge(s) at all, they lack any and all measurable mass as well.

Mass is, in effect, substance. Substance demands the presence of massive particles, typically atoms and their collective parts and nature(s). Virtual particles are again without mass. Actually, the total lack and/or total absence of any mass, so then the lack of any substance, may be the most unique and most difficult-to-define expression of nothingness. But, absolute lack of mass is mandatory, precisely for one very specific behavior…energetic movement.

Both photons do, and gravitons must travel at lightspeed. It is assumed that gluons do as well. Dr. Einstein's theories of relativity do so require that any mass(es) that reach lightspeed must and do become 'infinite.' So conversely, it makes perfect sense to begin with zero mass to (even be able to) travel at such a speed. Therefore, and again, massless virtual particles are seemingly the only entities permitted to do so.

What then becomes self-evident is that some mass (therefore, some substance) must be present to carry any electrical charge. Again, virtual particles or entities, having no mass, are absent any such charge(s). Conversely, those particles that have/exhibit mass carry an electrical charge… or not. Again, as above, neutrons in particular do not, due to their internal structures. But…

Electrical charges themselves may be either positive or negative. Net positive charges exist within all protons. Protons, like neutrons, are massive nuclear atomic particles with a plus one (+1) electrical charge each. Note that the constituent inner 'parts' of protons also consist of three quarks... structurally, much the same as do neutrons. However, of the quarks within protons, two are "up," carrying each their +2/3 charges, and one is "down," carrying its -1/3 charge. So the proton's electrical mathematics is also simple:

plus 2/3, plus 2/3 = 4/3 minus 1/3 = 3/3 = 1

Similar questions do therefore exist for protons as they did for neutrons. Question 1: How and why do protons congregate in atomic nuclei? Being charged each +1, they repel one another, as do all like-electrically charged entities. And that would explain the requirement for the strong nuclear force. Without repulsion at all, either between protons or between quarks within nuclear particles, such a constraining nuclear force would not be necessary. Then question 2: What then actually invites or initiates protons to congregate as they do in atomic nuclei in the first place?

Answers 1 and 2: "Likely extreme pressure and heat do." Again, these are subjects worth further discussion. (Refer to chapter 11, "The Significance of Threes"; chapter 12, "Particle Gender"; and chapter 13, "Dr. Sagan's Apple Pie.")

Negative one (-1) charges exist also within electrons. All non-ionized atoms are organized or reorganize themselves to contain the same number of protons and electrons...always. All atoms, therefore, that exist as the basic and organized units of all matter are electrically neutral. They do, same as neutrons do, and for similar reasons, display net zero electrical charge(s). Electrical "nothingness" is the rule at the atomic level. Atoms stripped of their electrons behave quite differently as ions. TBD at another time.

So then the electrically neutral math for atoms confirms their elemental order exactly and also by atomic number. For examples:

Atomic Number	Protons	Electrons	Net Charge	Element
1	+1	-1	= 0	hydrogen
2	+2	-2	= 0	helium
3	+3	-3	= 0	lithium
up to 92	+92	-92	= 0	uranium

Uranium, element 92, is the heaviest naturally occurring element on Earth. Hydrogen, being number one and the lightest, displays the identical electrical mathematics...net 0 (zero) electrical charge(s) in its elemental (so non-ionic) state.

Atoms (and neutrons) exist as electrical 'nothings' therefore...and each with an obvious purpose. Were they, atoms in particular, with any net charge at all, they would be repellant. Accordingly, either all matter, so all substance would be impossible, or another 'atomic (very) strong force' would be required...which, fortunately, it is not!

Conclusion: Both net electrical charges and mass may apparently go to 'nothings' and so to zero values, and both for very good reasons. What then about dimensions?

Any 'point' is conceptually and mathematically assigned zero dimensions. Points are assumed and are understood to express and to contain neither height nor width nor depth. This assumption is currently being challenged by String Theory. If it is difficult to assume dimensions where none are allowed to exist, consider again the logic: One must be able to cut anything, so any dimension or any mass (therefore, any temperature?), in half endlessly, and then <u>always</u> have <u>any</u> half remaining. (Refer to chapter 15, "Absence of Anything.")

<u>Zero</u> mass and <u>zero</u> dimension (and absolute zero temperature) may be more theoretical than actual? If so, how then can (zero mass) virtual particles or quanta travel at lightspeed? To perhaps resolve any contradiction, virtual entities must be regarded as being indivisible. Nonetheless…

If not then or not ever to zero and therefore to "nothingness," how small can anything(s) or can any measurements become? "Unconfirmed" would be a good answer, currently. There may be no limit to "smallness." Our mathematics, or more precisely, our measurements, may well be too coarse. Points must contain dimensions within dimensions therefore…as string theorists will agree! Theoretical physicist Max Planck, has calculated both a minimal (so nonzero) "Planck" dimension (1.6×10^{-35} meters) and a minimal "Planck" mass (2.176×10^{-8} kilograms). Logic must prevail that these 'minimums' are, like their 'maths' are, theoretical only. Masses or dimensions smaller may not matter (to/for mathematicians), but to logicians, they must exist. TBD. If "nothingness" is even plausible, then "anythingness" is contradictory. Both Planck minimums <u>are</u> <u>non-zero</u> nonetheless.

Reference is made back to an electrical charge…or not. For atomic particles, there currently exist permitted charges of plus one, zero, or minus one. Subatomic particles, more specifically quarks, exhibit fractional charges, and being the smallest bits of matter or substance currently known with mass, perhaps they do exhibit the 'finest' 1/3 charges of all? It seems fair to assume future cases, some unrealized example(s) where electrical charges of less than 1/3 may apply? These contentions as expressed may apply accordingly to both electrical charges and dimensions and masses as well.

A recent quote, authored by one Corey Boswell, seems appropriate:

> Science is full of zeros. Light has zero mass. Neutrons have zero charge. A point has zero length (actually, has zero dimensions). Those zeros might be unfamiliar, but they all follow a constant logic. All represent (actually, they all require) the absence of a certain quality: mass, electrical charge, distance, etc... (Parentheses added)

Unmentioned so far and above are potential absences of both motion (aka velocity) and time. Motion first...

Can there be <u>zero</u> motion, so <u>zero</u> movement? <u>Zero</u> velocity? Theoretically, yes. If light travels exactly at 299,792,458 meters per second (therefore 670,616,629 miles per hour) as it does, then <u>that</u> speed must be measured against <u>no</u> speed at all, which is zero meters per second (hence, zero miles per hour). But, can zero meters or do zero miles per whatever as above actually be found? Specifically, can 'no motion' (so zero velocity) actually exist?

It seems, as if from a cosmic perspective, so as if looking in at the universe from a constant point of reference outside, then zero motion does not exist. Even a constant point of observation, as above, may be moving. (Refer to chapter 9, "Mario's Champagne.") However, from a much more 'local' perspective, and assuming a reference to any one observer's static space-time (so conceptually, really, experiencing all things with reference to oneself), lightspeed, denoted by (small) c, does exist...therefore, no speed at all must exist as well...'relatively' speaking.

Thanks again to 'relativity,' aka Dr. Einstein: $E=mc^2$, and if so, by dividing each side by c^2, then m (mass) = E/c^2. This is to say: If energy (E) equals mass <u>times</u> lightspeed squared, then mass (m) equals energy <u>divided</u> by lightspeed squared. This rather simple algebra applied to a very complex equation produces a logical conclusion. But then, what about the notion of actual zero velocity? Note that c^2 is a fixed value, a fixed velocity of $\pm 9 \times 10^{16}$ or roughly ninety quadrillion meters squared per second squared. If ever attempting to compare mass to energy at zero velocity, the equation fails. This is to imply that (some) velocity is (must be) present at all times. (Refer to chapter 7, "$E=mc^2$.")

To compare energy to mass at zero velocity, both mass and energy must equal zero, and the equation becomes pointless. $E=mc^2$ itself essentially demands motion.

So then, take a moment for the specific mathematics of zero: Zero's concepts allegedly first occurred in/within Sumerian culture some 5000+ years ago. Such a time, currently referenced as ±3000 BC, coincides nicely with the budding Egyptian culture as well. The Egyptians, being well aware of geometry and of "Pi" (Refer to chapter 21, "Pi Is Irrational"), most likely were also well aware of zero. Nonetheless, and so much later…

The more specific and documented mathematical considerations of zero became evident in East Indian culture (AD ±600) then possibly Cambodian (AD ±700) then Chinese (AD ±800)…and then, perhaps lastly (actually, for the first time) in Western European cultures. "Zero" itself was first officially 'documented' by Italian mathematician Fibonacci in AD ±1200, relatively recently…actually, four-thousand-plus years <u>since</u> the Egyptians!

By almost all accounts in history, the simply logical and <u>conceptual understanding</u> of zero came first, and then its specific <u>mathematical applications</u> came along much later… so again, quite recently in human history.

Nonetheless: Cut off or lose a finger (or toe) and four remain. Lose two of either and three remain. But, cut off a hand (or a foot) and zero, or none, remain. Cro-Magnon people, some ±40,000 years ago, probably even *Homo erecti* individuals some hundreds of thousands of years ago, <u>well</u> before any Sumerians and/or Egyptians, must have been 'painfully' aware of such losses that resulted in <u>zero</u> fingers or toes!

Zero plus zero = 0. Zero less zero = 0. Zero times zero = 0. Zero times any number = 0. Any number less any same number = 0. So zero is indeed a number, and it exists as an <u>even</u> number as well. Zero occurs between plus one and minus one, both of which are odd numbers. Yet zero divided by two = 0…this is unlike any other even number when divided by 2. And these odd comparisons become more interesting when further considering division.

Again, $0/2 = 0$…understandably. "Zero divided by any number equals zero." So inversely, what can be said about any number divided by zero? Given $1/0$ or $29/0$ or any number$/0$, what is (are) the answer(s)? Three possibilities exist: (1) Any positive number when divided by zero might equal positive infinity. Or, conversely: (2) Any negative number when divided by zero might arguably then equal negative infinity. Positive infinity may make some (albeit, mathematical) sense, but negative infinity makes very little conceptual (or mathematical) sense at all. How to reconcile…?

The official and understood "PEMDAS" rules of mathematics state that: (3) <u>Any</u> division of <u>any</u> number or of <u>any</u> value by zero equals "NaN," which translates to "Not a Number." Therefore, per rule number 1 concerning division by zero, such division is "not allowed" because such computation(s) "has (have) no answer(s)." Such a rule as number 1 may be accepted but not exactly understood because...

Of rule number 2: "Any number divided by itself equals one." Recall that zero is a number. Therefore, if $1/1 = 1$ and $-1/-1 = 1$, then $n/n = 1$, and so $0/0$ should equal one...yes? But no, it does not. $0/0 \neq 1$. Rather $0/0 = $ NaN. Rule number 1 (that division by zero is not permitted) trumps rule number 2 (that any number divided by itself equals one). Seems unreasonable and unfair!

So $0/0 \neq 1$ is mathematically correct, but is also conceptually difficult to accept, let alone to explain...exactly. What actual value or number is "Not a Number"...exactly?

Back to the title and to the intentions of this chapter: How do the mathematical 'oddities' of zero apply to the conceptual notion of "Nothingness" and/or to the mathematical actuality of "Nothingness"? Does the logic agree with the science, so with the physics (aka FIZZICKS) in particular? Again, given that $E=mc^2$, and therefore that $m=E/c^2$, when and under what conditions can E or m equal zero? At what time might both energy and, therefore, mass have been totally absent? When did both $E = 0$ and $m = 0$ simultaneously...if ever? Obviously, there must be some time for E to equal $E=mc^2$.

However, such a condition must have logically existed just a 'moment' before 'our' time, $t = 0$. How possibly? It is not an easy concept, but might there have been some 'other time' before 'our time'? Might our $t = 0$ have been that moment between times? Is there, or better, was there a "differentiation" of time(s)? Hard to say, but such a possible condition would more easily explain our origin...our Big Bang.

Regardless, effectively at our $t + 1$, both E, presumably first, and then m, so at $t + 2$, burst into existence. So the differentiation of actual time(s) between $t = 0$ and $t + 1$ and $t + 2$ must be very small indeed. The notion again that our universe's "Bariogenesis" phase number 3, when primary atomic particles (the protons and neutrons) took shape, has been estimated to have occurred between 10^{-35} and 10^{-33} of the first second. So then the "Bang" itself, at $t = 0$ by definition, had to have itself existed between then 0 and 10^{-36} of the first second…etc. This sequence must have been critical.

Given any/all of the miniscule computations involved, whether 10^{-36} or even 10^{-360} of that moment, it still demands that $t = 0$ was in fact a time, when there was essentially no time, so nothing…<u>really</u>??

Recall that (small) c is our measurement of a very specific velocity, which is the speed at which (again) photons transmit electromagnetic energy (or energies) and at which speed gravitons must also transmit gravity. Both energy forms, electromagnetism and gravity, must be related. (Refer to chapter 18, "Absolute Zero," plus "Considerations," Chapters 18[A] and 18[B].) Either one's, either gravity's or EM's, reduction results in the other's increase or vice versa. So then to the mathematical comparisons…

Velocity (and c is <u>the</u> one under consideration) is specifically anything's measured change in any distance divided by the exact corresponding change in time it took to traverse that distance. Velocity equals Δ d divided by Δ t, where Δ indicates these changes. If confusing: A car or any vehicle traveling, say, at 60 <u>miles per hour</u> will have moved from point A to point B if there was an <u>exact</u> change of <u>one mile,</u> in <u>exactly one minute</u>, which was/is the <u>exact change</u> (read 'passage') of time. It took exactly 60 seconds to do so. Velocity is always expressed as a fraction: distance/time. And time, being always in the denominator, must always have a value, however brief it may be. Because again…

Division by zero, in any fraction, so <u>including any velocity,</u> is not permitted. Any distance divided by zero time produces "NaN," not a number. By implication then, at t=0, again, there was nothing to even be able to move nor to have been moved. Both the mathematics and the logic seem to confirm this to be true, that t=0 must be the ultimate statement of "nothingness." Therefore, what "something" must have been present at t-1 remains unexplained.

Neither actual nor conceptual (say, mathematical) explanations of velocity are permitted until t=1, and this is a requirement, no matter how small the measurement of said "1s" might be. Our universe must be and must have always been "in motion." This is whether we realize it or not! And, this assumes there was a time <u>before</u> when it, <u>our</u> time, was not...not even in existence!

What is remarkable and remains and so <u>is</u> and must be that: <u>All</u> of E (again, presumably energy first) and then all of mass had to have burst (<u>almost,</u> but not exactly simultaneously) into existence at a 'point' (wherever that was exactly) some 13,750,000,000 (Earth) years ago. The assumptions required may be overwhelming, but no better explanations exist.

<u>These</u> primary <u>assumptions</u> exist much like any legal assumption does. It's either 'guilty' (so it did happen that way) or 'not guilty' (so it did not happen that way). It is not (not 'legally,' anyway) 'innocent,' which would demand that it had to have happened some other way...so to result (somehow?) in the same way.

That had to have been 'it.' Our origin had to have occurred at the point of infinite (so still unmeasured/unmeasurable) <u>density</u> (so at dmax), which exhibited infinite <u>pressure</u> (so at Pmax) and resulted in (still) infinite heat, call it (still) unmeasurable temperature (so at Tmax). And so on...

Which 'max' actually occurred first may be irrelevant… or not. Was it density (of what?)? Or pressure? Or temperature? Which, in fact, is most likely? TBD. Regardless, this all had to have occurred at a point, potentially without dimension(s), so effectively 'nowhere.' Its timing had also to have been exactly at that moment when $t = 0$, so effectively 'never'! These implied initial conditions, represented by said <u>zero values,</u> may seem to present the ultimate contradictions, unless…

"Wherever ya go, there you are" is true. As stated in the Foreword hereto, it's a line from an entertaining science-fiction movie, so spoken by Peter Weller, the actor (aka "Buckaroo Bonzai"). Another line: "Ya gotta be somewhere," says much the same thing…and was/is the punch line of several humorous situations that qualify! Both quotes speak to the absolute and primary requirements of both space and time. Space zero or 'Sp=0' has never been a concept worth consideration. It, like division by zero, is not permitted! But $t = 0$ has been assumed because…

It has been assumed because there may be the possibility of some time before our time or t=-1. Let's refer to any such "time before our time" as τ, that's the lower-case t in/of the Greek alphabet. So it may have to be conceptually $t =$ our time, and $\tau =$ prior time. If possible, and if so, then:

1. $t = 0$ is the same as "t (existing time) starting" and…
2. $t = 0$ is the same as "prior time ending" or perhaps τmax?

Enough conjecture. However…

Both space and time must be older than 13,750,000,000 Earth years. End of story!

Further 'referential' circumstances are contained in the footnotes that follow…

Absolute "nothings" are very difficult, if not impossible, to prove. What is the "absence of anything," really? "Absence of Anything," chapter 15, herein, attempts to explain and when and if explained, there will necessarily be (and have been) "something(s)." Zero itself may in fact be a number, but what it stands for, exactly, is not for sure!

Chapter 19A

Consideration Number One Concerning "Nothings" That Result In "Somethings"

"Nothings," again, are described by zero values. These, as opposed to...

"Somethings." Somethings may be described with very big values, as in light-years and mega parsecs, for examples. And, the one very big value that essentially defines and/or explains almost all the others is c:

299,792,458 meters per second...obviously
well above zero

That, above, is light's speed, or more exactly, it is the exact velocity of photons in a vacuum...one very large number for one word, which exists as the one governing "something"...actually of most all of theoretical physics and cosmology, etc.

Lightspeed is the same for all radiant, electromagnetic energies. Why plural (energies)? Whether radiation itself (think of gamma rays and x-rays), or light and color (obviously), or microwaves (produced in ovens in most kitchens), or radio (or television) waves, or electrical energy waves…they all travel at this very same speed, same velocity, which again is:

299,792,458 meters per second, exactly when
in a vacuum

This is also exactly the same speed as is the flow of gravity. One may observe that some conditions, such as atmospheres and glass lenses and optical cables (and wires), both with resistance, slow down the flows of light (and electricity). In fact, lead (Pb) <u>stops</u> x-rays…all true. So <u>true</u> lightspeed is calculated 'in a vacuum.' But the very same gravity speed is the same in all situations. Nothing understood or known yet slows down gravity. No vacuum is required…for future discussion.

(Refer to chapter 8, "LightSpeed.") Would not the best constant again be <u>gravity</u> speed? Lightspeed is actually a maximum speed limit…which slows down when not in a vacuum. Gravity speed seems to be <u>the</u> best actual constant… anywhere and anytime. Its speed appears to be constant, whether in a vacuum or not.

We all seem to accept what a second is…a measure of time. A second is currently defined as the duration required for a cesium atom to 'cycle,' that is essentially to vibrate roughly 9.2 billion times. There are 60 seconds in a minute, so 3,600 of the same 'durations' in an hour. Therefore, both minutes and hours (and days and weeks, et al.) are really designated amounts of seconds…just as divisions of time it currently takes the Earth to rotate once (which currently is 86,400 seconds).

Do a bit more math and compute how many vibrations a cesium atom experiences in a year, or a century! These are very, <u>very</u> big numbers!!

Keeping the science(s) all together, again, is lightspeed = (small) c, and c is expressed per second, which again is the globally understood measure of time that always results in the same number of meters traversed. So then what is a meter?

It's all too easy to elude the "What's a meter?" question just asked with the answers of: "A meter is exactly 39.37 inches" or, more exactly, "3.2808 feet." But these are not answers. Inches and feet are really other measurements of… what? And if 'of what,' why use 12 (inches in a foot) or 3 (feet in a yard), let alone 1,760 (yards in a mile)? Math with 3s or 12s or 1,760s is awkward. Meters (and so metrics) utilize the math of 10s, which is simple. Understood. Still, what is a meter??

The best and first original explanation (theoretical then and still) was an alleged measurement equal to 1/10,000,000 (that's exactly one ten-millionth) of the distance between the Earth's equator and its north pole. Said distance was originally the alleged measurement of such a line drawn through Paris. That was back in 1794. Never mind, it was not then known exactly where either the equator or the poles (either one) actually were. Accordingly, the concept and said measurement was decidedly both theoretical and very "French." Regardless…

The calculation produced a result, however accurate or not, which resulted in an arbitrary length, which defined a meter (also arbitrary). And that result, that computed distance, was inscribed and was so marked on an iron rod that was stored…in Paris. And this would be the reason this measurement is properly spelled (by the French, certainly) as the 'metre'!

In 1875, the original rod was replaced with a more durable platinum-alloy rod, which was properly remarked with the original 'metre.' This now official rod and 'official' measure were and remain stored in the International Bureau of Weights and Measures facility, also near Paris, in St. Cloud…so close enough. It all, both the rod and several other standard measures and the Bureau, remain very much in France, as they have been now for roughly 229+ years.

And then, in the interim, light's speed has been measured several times in essentially similar ways. Mirrored tubes were constructed of exactly xx meters length(s) and 'vacated' of atmosphere, and with one clock coordinated with the light source, which were switched on and off simultaneously, the light beam traversed twice (as reflected) the xx distance(s)… which have the same, average results, which were and confirmed light's speed to be:

299,792,458 meters per second, confirmed

The math was relatively simple as long as all the clocks and light sources were properly coordinated and the essential vacuums were maintained. But even the original platinum rod might be subject to change(s) or variation(s) over time. If subjected to heat, it could (would?) lengthen, or if rendered very cold, it might shorten. In fact, its markings themselves would and <u>did</u> include some widths of their own. So to be sure of no errors over (any) time(s) or conditions or temperatures…

The original lightspeed fraction: 299,792,458 meters per second was inverted to read 1 second = 299,792,458 meters. The ratio of distance over time was essentially reversed to create a new standard measurement that reads:

One meter is the distance that light travels
in 1/299,792,458[th] of a second in a vacuum.

Doing the math again, an accepted meter is the distance a photon (therefore of any/all electromagnetic energy) travels in 3.336×10^{-6} of a second when in a vacuum. Additionally, and since...

A meter has been (perhaps, unnecessarily?) measured and so additionally defined as the same, very exact width of exactly 1,650,763.73 wave lengths of a very specific red-orange light. Therefore, the conclusions:

1. A second is defined as the duration of time required for a cesium atom to 'cycle,' so in effect to vibrate exactly 9,192,631,770 times.
2. A meter is equal to:
    a. One/ten-millionth of the distance (recall, through Paris!) between the Earth's equator and its north pole. And therefore...
    b. The distance between two marks on a platinum rod in St. Cloud, France. More specifically...
    c. The distance light (in a vacuum) travels $1/299,792,458^{\text{th}}$ of a second. And most specifically...
    d. The width of 1,650,763.73 red-orange light waves (assumed also as measured in a vacuum).

Some fault(s) can be found in any or all of the above. How were (or are) perfect vacuums presented and maintained? Lightspeed verifications have been made via reflections between vehicles and objects in space. Laser beams have been bounced off the moon...better yet? And one may go on and on seeking confirmation.

But now, one is really able to say "somethings" rather than "nothings." Zeros are relatively 'easy.' It's the big numbers that can be confusing. We've established and agreed upon c...

hope it's right! Whatever it is, Dr. Einstein has confirmed that lightspeed in a vacuum must be and is equal to gravity speed in…anywhere. So, if photons (again, confirmed) transmit light, and if gravitons (still theorized) transmit gravity, this, their speeds, cannot be just coincidentally the same. They must be related therefore.

Zero motion versus ±186,282 miles (same as ±300 million meters) per second is a big difference. All sorts of interesting 'events' and 'behaviors' occur most specifically with matter and so with mass(es) at either of these extreme values.

"Somethings," much the same as "nothings," do require all sorts of explanations, and do lead to all sorts of additional and fascinating conclusions.

Chapter 19B

Consideration Number Two
Concerning "Nothings"

Zeros in physics (and in most any science) express zero values...obviously. 0s = 0s. Zero values in physics are consistent in the very singular sense that they all (and always must) express the total absence(s) of some specific characteristic or of some quality or of some condition...or of some other metric, however described or measured. So, consider again our beginning...our Big Bang.

It was, by definition, our original time, when that very first time t = 0. Imagine the moment, which, by deduction, had to have occurred just prior, when there were:

1. no temperature(s) at all,
2. no electrical charge(s) at all,
3. no mass(es),
4. no dimensions, therefore
5. nothing(s) physical at all.

Therefore, with 'nothings physical,' so 'nothings actual,' there were no starting nor ending points and nothing(s) to move. So, there was no motion at all. Given either that motion creates heat, or that heat results in (and actually requires) motion, three more absences become self-evident just prior to t = 0. There also were:

6. no motion(s), so
7. zero velocity (of anything), therefore
8. no energy (nor energies), which confirms…nothing (no pun intended!)

Back to 1. no temperature(s) at all.

And an allegorical chain of events, better, a chain of nonevents as above, must be complete. Any breaks in such a chain would render an end to any circumstance(s) that then did occur at t = 0.

Noticeably, of the eight original zero values above, and in most any 'order' or in either 'direction' they might be considered, each one implies, or actually demands, the seven others. Absent any one of the eight above, it is difficult, if not impossible, to assume or to argue the presence(s) of any of the other seven. Stated more succinctly:

> "Nothingness" seems to be unavoidable and/or inevitable when any one original, physical quality or any one original, physical condition is absent.

Thus, the above statement itself is coincidentally (perhaps eerily?) consistent again with the mathematics of zero, wherein:

> "Anything times zero = zero."

The statements above, both of them, exclusively ignore the one other condition that occurs always in the presence of mass…and that is <u>gravity</u>. As important and ultimately omnipresent as is gravity, it/gravity exists only in the presence of mass. And, mass was not apparently present just prior to t = 0. Therefore…

Gravity, as understood currently, may be ignored as an energy form when discussing the presence(s), or not, of the actual, original conditions exactly at t = 0. However, gravity of perhaps of some 'prior' existence of some prior mass must have been a necessary 'prior' condition…again, prior to t = 0. It's a bit confusing, but a requirement seems to exist nonetheless that:

"Gravity exists only in the presence of mass."

Not only conceptually, but then actually, what <u>other</u> prior and <u>necessary</u> (in effect, beyond physical) <u>qualities</u> and so <u>necessary</u> prior <u>conditions</u> must be and so <u>must have been</u> present prior to t = 0? Answers:

A. Space
B. Time

If again, <u>our</u> Big Bang occurred at <u>our</u> time zero (t=0), then <u>that</u> time in fact <u>did</u> exist roughly 13,750,000,000 Earth years ago. And then two, obvious questions apply always:

1. What was present? And then accordingly…
2. What was absent…so what did or did not exist roughly 13,750,000,001 years ago?

And the third and fourth obvious questions become unavoidable:

3. What could have been that 'time' before our time? And lastly…
4. Where? In what pre-existing space did this occur?

Items 1 through 5 at the beginning of this chapter pretty well sum up what were <u>not</u> there prior to t = 0…but they may not explain the total of any and all absences. (Reference herein is made back again to chapter 15, "The Absence of Anything.") So, from another perspective…

Another alternative example or altered state of one's given concept of "nothings" might again compare to one's understanding(s) of a 'point.' On the theoretical aspect, a point is denied any dimension(s)…no height, no width, and no depth. However, from the actual (so String Theory) aspect, that very same, so any point has 'wrapped up' (so to speak) inside its own 'self' countless dimensions within dimensions within dimensions…as previously implied. (References here are made to chapter 17, "Dimensional Proportionality," and again to chapter 15, "The Absence of Anything"…both for added consideration.)

"Nothing" itself is much like a 'hole' and/or a conceptual 'vacuum' in which anything <u>can</u> occur. Witness again the apparent presence first, then absence (the 'hole' again), then the return of gravity. And this concept is and has been previously conditioned upon there always having been a 'place' (so a <u>space</u>) and therefore a circumstance (so a <u>time</u>) for anything at all <u>to occur</u>.

Given again that: Both space and time are primary and <u>necessary requirements</u>, they are therefore <u>requisites</u>. Neither of their absences are permitted. If they were to be, somehow and in some effect quantifiable, then:

$$\text{Both Space}/0 = \text{NaN}$$
$$\text{And Time}/0 = \text{NaN}$$

Enough for now of "nothings." From yet another aspect entirely, what of "somethings"? If the above have dealt with the 'negative sides' of all things, consider now what must exist on the 'positive sides.' It must be considered the 'plus side' where too much of anything exists. Wherever and whenever (space and time again!) just enough or too much mass concentrates into a small enough space, which would be again considered small enough to be a point, and therefore enough density is created to result in enough pressure to create enough temperature, then a 'singularity' occurs by definition. Again, such a singularity is defined by a lack of, and therefore by the theoretical absence of any dimension(s)...which explains, or better, which describes only that one absence... of dimension(s).

What then is also absent, simultaneously, if just for that same moment, must also be the absence of energy, which is herein and hereby described as a "similarity." So the two technical absences are distinguishable as:

1. The lack of dimension(s), which defines the "singularity," as compared to:
2. The lack of energy, which must define the "similarity."

The (any) similarity is such that it must describe the momentary transition of (essentially) maximum gravity to the existence of essentially (actually?) maximum heat. What actually happens in essence in between defines the 'similarity' itself. And…

This seemingly confirms again the observation that:

Gravity and heat are inversely proportional.

Which is the same to say that:

"When gravity is absolute, so that it is at its maximum, there can be no heat. And so, when heat is at its maximum (also, potentially its 'absolute'), there can be no gravity."

If any relationship of any quality or condition is, in fact directly inverse, then at either extreme, the absence of one or the other must be evident. Reflecting back then to chapter 18, "Absolute Zero," the similarity between and for the transition(s) between photons and gravitons cannot be ignored! (Or if ignored, then ignorance must be present!) And further…

These existences (of both minimum dimensions and maximum temperatures) seem to be the identical models for larger stars in our known universe that do collapse, and therefore for those that have collapsed to result in supernova (so 'large-star') explosions that have been observed. Said supernoval explosions, in effect, must occur in two directions:

1. The first direction is evidenced by a massive shock wave that travels outward as a radiant energy surge or 'shockwave' that creates all sorts of naturally occurring (i.e. non-man-made) elements heavier

than number 26, iron (Fe)…as has been previously discussed.

2. The <u>second</u> (and simultaneous) <u>direction</u> must be an (identical?) energy impulse (better, 'compulse'?) that travels <u>inward</u>, so more as a convergent energy pulse that, when sufficient, forms points that must be the centers of black holes. How else can their creation(s) be explained?

Curiously, that <u>impulse</u> or <u>in</u>ward energy force sounds suspiciously the same as gravity…but cannot be the same. TBD. Nonetheless: "Every action (the first/outward) results in an equal and opposite reaction."

Black holes both require and exhibit some defined limits, and these limits must be similar to absolute zero (Refer to "Absolute Zero," chapter 18.) and absolute temperature. But still, and however they may actually be measured, these limits may still fall just short of "nothings" on the downsides, and just short of "everythings," so better, just short of "maximums" on the upsides. What or who then decides these limits?

The mathematics, specifically the currently agreed-upon limits, have been established by one Max Karl Ernst Ludwig Planck…the <u>one</u> theoretical physicist with <u>four</u> first (and second and third) names! Max Planck was born in 1858 in Germany, twenty-one years before Einstein, and died in 1947, then years before, so both physicists were busy during roughly their same time(s). Planck originated quantum theory and was rewarded with a Nobel Prize in 1918 for doing so. Einstein was awarded his Nobel Prize for his Theories of Relativity in 1922. Both almost simultaneously established the current two 'pillars' of physics: (1) Relativity Theory and (2) Quantum Mechanics.

If Einstein's lasting contribution was/is his most famous equation: $E=mc^2$, then Planck's contributions were his two minimum and one maximum measurements, often if not imperfectly referred to as "Planck's constants":

1. The minimum recognizable and therefore measurable distance of "Planck length" (Lp) is 1.62 × 10^{-35} meters. Note again the proton, a primary nuclear particle, measures ±1.73 × 10^{-15} meters in diameter...therefore, 10^{20} or some 'sextillion' times larger. Planck's minimum distance is astonishingly small, but is non-zero just the same.

2. The minimum recognizable and therefore measurable moment of time (Tp) is 10^{-43} seconds... again astonishingly small "Planck time," but also some time.

The above two minimums are so stated because any lesser values somehow "don't exist under the current law(s) of physics." There also do not exist, as there logically cannot exist, many maximum lengths, distances, or time(s)...as opposed to:

3. The maximum Planck temperature of 1.42 × 10^{32} Kelvin, which again, given the existing laws of physics, prohibits even 1.43 × 10^{32} or any 10^{33} measurements of heat at all. And accordingly, temperature must have a range, absolute zero (0 K) to 1.43 × 10^{32} K, thus indicating both the absence of all and the presence of all heat. Is this "Tmax"?

There may be some even more exact measurement(s) (so other nonzero numbers) to attempt to define the smallest/

minimum mass(es), but Mr. Planck apparently did not as clearly establish this/such a value. His <u>stated</u> value of 2.18×10^{-5} grams is obviously not (so cannot possibly be) the smallest mass. Since Mr. Planck's time, the mass of an up-quark, for example, has been calculated to be 3.922×10^{-27} grams, so then the smallest unit of mass yet identified... and is itself some twenty-two times smaller. This restates the obvious: "Any mass (and therefore any measurement) can be divided by one-half...endlessly."

Thank you, Mr. Planck, but science and mathematics have and will continue to improve themselves!

Referring back to the centers of black holes: Where and when they exist, our current laws of both physics and of mathematics fail. So the details of their actual qualities or actual conditions and then actual measurements remain unknown...but perhaps are not actually unknow<u>able</u> at some future date. Therefore, both the physics and mathematics for our Big Bang must—however on a much grander scale—be also both identical and unknowable. But, while black holes may recycle stars, Big Bangs, both ours and potentially others, must either:

1. recycle universes on the same dimension(s), or
2. create universes on or in a reduced dimension.

Such appear to be the positive and negatives so involved. Were Mr. Planck still alive, it would be most interesting to hear his perspective.

Chapter 20

The Opposite Of Entropy Is <u>Contropy</u>

If a picture is worth a thousand words, then graphic representations or columnar comparisons, as below, with more or less one hundred words, may do as well…hopefully:

Conditions and Ultimate Parameters

Parameters	Entropy	Contropy
1. Locations	everywhere	anywhere
2. Dimension	infinitely large	infinitely small
3. Temperature	to -273°C., or…0°K.	infinitely hot, or…Tmax
4. Pressure	0	infinite density
5. Organization	none/minimal	specific/exact
6. Proportion	larger	smaller
7. Structure	simpler	more complex
8. Architecture/ End Point	chaos	singularity

So then:

Parameters	Entropy	Contropy
9. Direction	outward—radiant	inward—centric
10. Heat is…	lost—diluted	created—increased
11. Density + pressure are:	reduced	increased
12. Time is…	forward only	unknown (TBD)
13. Because of/ due to:	Thermo Law #2	(our) universal gravitation

So that perceptions are: Ultimate Observations:

Parameters	Entropy	Contropy
14. Understanding is…	Lacking and dubious	specific/definable (to a point)
15. Information is therefore:	most general and confusing	more specific and understandable
16. Energetic expression is:	None	maximum

Therefore and as above, with less than one hundred descriptive words:

1. Entropy will proceed to demonstrate at its maximum:
   a. a state of disorder and disorganization, that
   b. lacks specific understanding, and is
   c. generally without information and confusing, and requires
   d. maximum space, that expresses
   e. the specific absence of heat energy, and results in
   f. chaos therefore.

300

This, as compared to:

2. <u>Contropy</u> that will proceed to create:
    a. a state of specificity and order, that
    b. demands specific structure, and is
    c. generally understandable, and consumes
    d. minimal (if any?) space, which expresses
    e. the specific presence of maximum gravitation, and ultimately results in
    f. a singularity.

Entropy is easier to conceptualize and makes sense in the setting and assumption of an expanding universe, but again is difficult to define if and when at its maximum. How 'disorganized' can 'disorganization' be? How cold is 0° Kelvin...really? When is or when can entropy be 100 percent...if ever?

Contropy is easier to explain logically, and may be easier to compute mathematically...that is, until a singularity results. But, a singularity does define a specific end-point, which is a specific result. This, the creation of a singularity, seems to contradict the reasoning (of D'Abro and others) that increasing entropy (i.e., <u>less</u> organization) is actually <u>more</u> predictable, therefore more provable, given universal expansion. And yet, stars continue to be created at the same time as our universe is expanding currently. How long these two processes can or will continue to proceed in tandem would be very hard to know. When, if ever, will star formation end? Also and again, "if ever?"

The key to understanding our current 'place,' now and within our current circumstance, is that both processes, both entropy and contropy, are occurring simultaneously. As long as our universe continues to expand, and when measured at any point in any foreseeable future time, less temperature and less density (and perhaps less density of information) and therefore increased entropy would be logical to assume. As any space expands with a fixed amount of 'stuff' (matter and energy) within, then the net amount of said 'stuff' that ever has been will be diluted at any point or in any location as specifically measured at any future time. And yet...

Our interstellar, say intergalactic space is definitely not empty. On its grand scale, and when considering all the existing galaxies and stars and planets and moons and asteroids and comets and 'dust' particles, et al., our universe is surprisingly still full of 'stuff' that are objects, which consist of matter, which have/express mass(es) therefore. Or, there exists something else that also demonstrates mass, and that has yet to have been identified, but that responds to or displays evidence of gravitation, nonetheless. Dark energy? (Refer to chapter 15, "The Absence of Anything.")

Gravitation exists wherever (and whenever) mass is present. Gravity is the measuring method that explains the current behavior of our observable universe, therefore. Dark matter? (Refer again to chapter 15.) Something, some more specific mass, is missing as well. Back to the topic of contropy...

Gravitational attraction between any and all 'bits' of matter (no matter how infinitesimal), and between any and all 'parts' (however small), and between any and all celestial 'objects' (be they stars or planets or moons again), and between all grander 'units' or 'organizations' (be they solar systems and/or galaxies), said total of all attraction(s) need be considered. Even when it, the total, is properly considered, there is not enough. There is not nearly enough mass that is visibly or even predictably accountable or explainable within our known universe that is enough to explain its observed gravitational behavior. Stated more simply...

Considering all the collective 'stuff' that is identifiable as 'matter,' and therefore that displays mass within our known universe, roughly 95 percent of what ought to be present is <u>missing</u>. How is this possible?

As restated then in a positive aspect: What <u>is</u> known to be here and to be present and that <u>is</u> evident and exists with measurable mass currently represents only ±5% (that is five percent) of what is necessary to explain the current behavior of our universe. And...

What is not (yet) observed and what has not (yet) been measured are theorized to be either of the following:

1. <u>Dark</u> (unseen and, to date, unidentified) <u>matter</u>, which is currently thought to represent ±25 percent of the total <u>mass</u> that is necessary. And...

2. <u>Dark</u> (again, to date, unidentified) <u>energy</u>, which is thought to represent 70 percent of the total <u>behavior</u> that is necessary.

So 95 percent, with 25 percent being attributed to dark matter or dark mass plus 70 percent being attributed to dark energy, is missing...as in, is currently unidentified and unexplained.

This is surely the most evident gap in our current understanding of our universe. What might be dark matter and dark energy, therefore?

Volumes have been written to again explain and to even identify these 'dark' mysteries. This one—this writing, this volume—proposes two alternative theories only...one each for the missing matter and missing energy:

1. A potential -1 (negative one) dimension (that of "Tim's" existence, in its analogy only) may well explain the missing matter. And...
2. A potential +1 (plus one) dimension (that of "Herb's" existence, also in its analogy only) may explain the missing energy.

While these considered and alternative dimensions may themselves be regarded as 'fanciful' or 'fictional,' neither their logic nor their rationale should simply be ignored. Again:

1. What explains and/or what occurs at a singularity? What happens to all the mass (of a very large collapsing star) that is condensed to a (theoretical) point in/at such a singularity? Is it, 100 percent of that mass converted therein to energy? And if so, where does that enormous amount of energy exist, and in what form does it exist subsequently, either within the singularity or elsewhere?
2. What can explain the original structure or original, collective mass that seemingly had to contract and then had to have collapsed in the first place to explain our Big Bang?

What was (and what is therefore) left, and that has been explained to date as a result of our Big Bang, is specifically that ±5 percent only that we can currently identify. All the rest, so ±95 percent of the total that explains what we/our sciences cannot, remains 'dark' and unexplained therefore. So...

Why should we (should anyone) be concerned??

It is best kept in mind, again and again, that the sum total of all gravity, so then of all attractive gravitation must exist to maintain the observed structure of our (and of any other?)

universe. And it, this said 'attraction,' must exist essentially as does time. That is to say that both must exist:

1. simultaneously, and
2. everywhere, and
3. absolutely.

Both our gravitation and our time must exist in all spaces (therefore in all places) and at whatever distances that may exist (or that may be assumed) in all said space(s). Gravitation specifically may vary only in the relative magnitude (of its force) and so the relevance therefore (of its attraction) as may vary with the difference of separations (so the distances between) any and all masses therein. This as opposed to...

Time, which is constant at any and all distances, but may vary (so must be inversely proportional) to and with velocity. So gravity, being inversely proportional with both distance and temperature, and with time, being inversely proportional with velocity, must have some yet-to-be-discovered commonality (or commonalities)...yet to be considered in *Fizzicks 201* and perhaps *301* and beyond. Keeping in mind that...

If inter-dimensional (Tim's) and extra-dimensional (Herb's) universes do exist, both their proportional times and proportional gravities must be in evidence and must apply to "Me" and to "Them," just the same as they do and must to us. (Refer to chapter 17, "Dimensional Proportionality.") It appears that some form of very special relativity remains to be explained and then to be understood.

There apparently is still plenty of 'science' yet even to be considered...and definitely yet to be found. And when considered and where found, it, the 'new' science, need be both defined and explained. Such explanations will surely have to include new measurements and currently unimaginable new quantifications and more theoretical mathematics.

We, *Homo sapiens*, and our evident human intelligence, however currently measured or demonstrated or itself explained, must obviously be currently explained as being (and as having been) 'rather primitive' and 'demonstrably naïve,' again now or at most any other prior point in our time. Recall that our ancestors, who preceded us by milliseconds only when measured in/upon the grand scale of time, worshipped the Sun and our Moon (and could not explain either). Milliseconds later, we can and have. So then what might happen in another minute or hour or so of cosmological time…if we can manage to exist that long? And, in addition…

Our very human brains have a very limited capacity and are troubled by opinions and perceptions, not always being accurate. Human brains also operate at a very limited (slow in comparison) speed. Even our nervous impulses, and so our actions and reactions occur, or better, are transmitted (within and by our neurons), and said speed has been measured at ±30 meters per second. However…

Artificial intelligence (think, machine learning) is not so limited. In fact, its (AI's) capacity (free of both opinion and altered perception) is essentially unlimited. Machines don't 'forget,' nor do they 'misspell' nor 'miscompute' either. And they can think very closely to lightspeed, so at nearly three hundred million meters per second…about ten million times faster than can we. Ergo: Do not enter a quick-draw competition or a Jeopardy game with/against a future computer or robot!

Artificial intelligence is now and will increasingly be our educator…that is, if we design and manage it properly. New subjects of and upon our current and future relationship(s) with intelligent machines abound.

It may be expressed, possibly, as a rule or even <u>the</u> rule (?) of universal balance that: <u>Entropy</u> proceeds to enhance and so to explain larger dimensions, while simultaneously, <u>Contropy</u> proceeds to create and to define the smaller ones... and both, in combination, produce the dimensional reality in which we exist.

How to confirm or to deny this? Consider two pertinent rationales:

1. "The <u>absence</u> of evidence <u>does not</u> prove the evidence of <u>absence</u>." However...
2. "The <u>presence</u> of ignorance <u>does indeed</u> imply the ignorance of <u>presence</u>."

So then again to the subject of this chapter: Contropy. What <u>does</u> happen, and so what <u>must</u> occur at a singularity, given a singularity must exhibit maximum gravity? Is maximum gravity enough to explain it all? What of a 'similarity,' as will be proposed in subsequent chapters...?

The opposite of entropy obviously exists, as it must, to properly explain both the mathematics and the current function(s) of our universe. Entropy and contropy occur simultaneously in our currently dynamic universe.

Chapter 20A

The Opposite Of Entropy,
Part One: The Science

As it is at the same time obvious and useful and educational, my bookcase presents itself daily just to the right of my desk. Its bottom shelf number 1 is messy…so it is essentially <u>disorganized</u>. It contains 'stuff,' as in various notes, articles, copies, supplies, assorted envelopes, and mailers, etc., that do not have any other assigned places. This, my shelf number 1, is <u>not</u> <u>organized</u> and may be regarded as "<u>more</u> <u>entropic,</u>" therefore.

Shelf number 2, just above, is specifically <u>organized,</u> better than most (of the three others just above it), and so much better than number 1 just below. Shelf number 2's contents are limited to my rudimentary collection of dictionaries. They are organized chronologically from the turn of the prior (nineteenth) century to the present. Included also are three thesauruses (thesaurusi?), also in dictionary form, plus several volumes that deal specifically with common errors and misuses of the English language, so then with the proper uses of English words. Shelf number 2 is much <u>more</u> <u>organized</u> and may be regarded as "<u>less</u> <u>entropic,</u>" therefore.

Were any scientist trained generally in thermodynamics or, more specifically, in wave theory, where he or she to read

the two opening paragraphs above, he/she might disagree...
possibly disapprove. Whether or not my shelves' organizations
do or do not demonstrate entropy is less a matter of opinion
and more a matter of a convenient analogy. If I were to be
equated as an 'organizing force' and then whether my efforts
were 'applied' or not, the analogy would fit. However, based
upon the original intentions, or better, the original conceptions
of the word *entropy*, it might not.

Back to an original reference text: *The Evolution of
Scientific Thought* by A. d'Abro (1927). "Entropy" was
then first described as "...the ratio of (a) quantity of heat to
temperature" (page 203), and is a very vague and inadequate
definition to be sure.

The basic premise of this early definition was: Absent
of any outside forces (so with no additional supply, nor
transferring of heat and with no increased density), heat
within any system always degrades and reduces to and toward
cold...and in that 'direction,' hot to cold only. So then, over
time...

Any object in any system that is hot cannot, by itself,
become hotter. Rather, it can and will, again by itself, become
colder over time. Again, per Mr. d'Abro: "Heat cannot follow
unaided from a colder to a hotter body, but tends invariably
to seek lower levels of temperature." Such 'direction' of hot
to cold is expressed as being one way and is absolute. And
further...

It is also in the same 'direction' and in the same sense
as the past flows (through the present) to the future. So, the
direction of heat became first synonymous with:

1. the "Principal of Degradation of Energy." And
   then...

The process by which heat, primarily, and then also by which any other form of radiant energy subsequently degraded or diminished, was described via:

2. the "Principal of Entropy," which was seen to proceed in the same direction as was the expansion or degradation, so then the increased disorganization(s) of any system over time, which then required:
3. the "Arrow of Time" also to proceed forward only and so by conclusion was confirmed...
4. the "Irreversibility of Time."

Viewing these assumptions in their reverse order, heat entropy became specifically:

5. the "Second Law of Thermodynamics."

The deductions and assumptions above may seem to be a bit overwhelming at first. Most anyone and everyone knows that a bowl of chicken noodle soup or that a cup of cappuccino will cool off by themselves, and that they cannot reheat or somehow become hotter if unaided. True, ice cubes in an iced tea (or a frozen cappuccino) will melt, so they get hotter indoors, thanks to central heating. And, as they will, so will snow melt outdoors due to the Sun (ambient atmospheric temperature). Oily rags if stacked tightly can also combust (so gain heat) spontaneously, and so chemical reactions (spontaneous combustion is one) do occur...but all such 'heat increasing' events occur with extra, exothermic, and/or chemical help. However...

A stack of wood in a fireplace may be ignited and will itself burn to produce heat, smoke, and ash, but cannot unburn itself. And the same wood cannot reconstruct itself...nor return to the wood pile! Thermodynamic systems proceed to degrade themselves via the loss of temperature (better, their loss of heat energy) but unaided and by themselves and in the same forward direction via <u>entropy</u>, just the same as do all events move forward from the past (also unaided) into the future via <u>time</u>.

Understandably, these concepts were at first considered to be uniquely cerebral, but have since become quite obvious and are now easily explainable and observed. So then the concept of entropy, in the original 'thermodynamic' sense, as being the loss of temperature, again the loss of heat energy over time, is perfectly logical. However, applying then the same logic and deduction to <u>probability (or probabilities)</u> over time required another 'step' intellectually. Considering then our initial universal state, the Big Bang, to be totally <u>improbable</u>, all subsequent outcomes become <u>more probable</u>, therefore. And one can compare our primal, very high temperature and very improbable universal event...again, the Big Bang, accordingly...

To explain...

Nothing currently imaginable would be possibly more improbable than "a point of infinite pressure and temperature" that simply "occurred somewhere (still, logically at the original center of our universe) in space" and at "approximately 13.75 billion years ago," and exactly "when time began," et cetera, et cetera, et cetera. Again, the probability of such an event to have happened exactly when and where it did or to reoccur at any future time is essentially zero. Further...

Said improbability is compounded by the concept that perhaps space itself did not (just as time did not) exist prior to this event...a future discussion of an apparent impossibility.

However...

If we may consider this, our "Big Bang" event only, and ignore, for the moment, that 'nothing' may have been in existence just prior to its occurrence, then what was the event exactly? It is currently thought to have been:

1. A dimensionless <u>point</u> (discussed in chapter 2 "Points" and elsewhere)
2. Of immeasurably high temperature (so infinitely <u>hot</u> as discussed below), and...
3. Of an immeasurably high pressure (so infinitely <u>dense</u>), wherein...
4. All energies (of the four currently understood) were <u>contained,</u> and...
5. All matter (also as currently understood) was essentially <u>present,</u> and all together they...
6. Existed <u>spontaneously</u> and <u>instantaneously,</u> and at...
7. Exactly <u>time</u> <u>Zero</u>!

Again, such an event must be as totally improbable as one may define, let alone as one may imagine.

But then, starting at time t0, so at the very first instance of our current universal time, the most perfectly organized point or 'structure' of all time, therefore the most improbable, existed nonetheless as our universe's singularity. And it is and has been expanding, cooling, and so increasing its size and entropy ever since.

Universal expansion, again from the <u>least</u> probable but also the <u>most</u> complex and <u>most</u> organized point imaginable to a forward and <u>less</u> complex and <u>less</u> organized status is therefore the exact expression of entropy. Whether viewed from the standpoint of temperature, or size, or density, or complexity, or kinetic energy, or of whatever the original conditions may have been, all standpoints of whatever are degrading and are diffusing and are becoming less specific and <u>more</u> probably into what will be <u>most</u> probably chaos, which will result in total <u>dis</u>organization ultimately…when considered upon the grand, universal scale.

Carried to the extreme and over time, our universe will be, or better, is currently theorized to most probably become, as compared to items 1 through 7 above at inception, as follow:

1. <u>infinitely large</u>, and…
2. of zero temperature (so <u>absolutely cold</u>), and…
3. of zero, unmeasurable density (so <u>pressureless</u>), and…
4. with massive, <u>total reassignment</u> and/or dilution of energies (to be discussed), with…
5. all matter <u>dispersed</u> and in a condition of…
6. either <u>minimum order</u> or <u>maximum disorder</u> (however to be measured or observed), and then at…
7. most possibly but still relatively, what will be the <u>end of time,</u> when zero structure and total chaos exist?

But, the seven outcomes above, with the possible exception of number 7 itself (TBD), are considered to be <u>highly probable</u>, so essentially inevitable, as our universal behavior is currently understood.

So again, considering item number 2, the temperature as above: The overall first theorized, currently observed, and now confirmed transition of heat from the initial temperature (again estimated) at 10^{32} + degrees Kelvin, has since degraded, dispersed, and cooled to the currently measured and confirmed cosmic microwave background (CMB) temperature of 2.7 degrees Kelvin or…properly expressed currently as 2.75×10^0 degrees Kelvin. So the current average temperature of our universe has been generally redistributed and/or has reassigned very nearly 10^{32} degrees Kelvin or an enormous measured amount of initial heat energy over its plus or minus 13.75 billion years of time to date. And it has done so via entropy. And the second Law of Thermodynamics, so then again of entropy, has been confirmed…as the result of universal expansion.

Over this same time and as our universe expands (as it is expanding), its probable condition at any future time(s) say at t2 or t3 etc., will be colder than at any present time, say t1 (or at any prior times back to t0).

So then "coldness" is more probable than "hotness." Stellar objects such as stars and galaxies and pulsars, so then any and all structures that create heat are, by definition, less probable, and so then may be <u>im</u>probable going forward. However…

They, all the heat-generating structures above, do currently exist, so something is simultaneously counteracting entropy and must exist in an extant "Opposite of Entropy," which is the title of this chapter.

Considering the obvious opposite(s) therefore: The more specific organizations of our space, our cosmos (again stars, galaxies, pulsars, etc.) are evidently less probable so more improbable, but at this, the same time, are extant. They currently <u>do</u> <u>exist</u> (period). But probably and ultimately, in time, they too will degrade and become less organized so more probably via entropy. If confusing…

Again, and in closing this Part One, per Mr. d'Abro's rationale (all with parentheses added):

We may say that natural phenomena present a unidirectional sense of advance, passing from states of lesser probability (so of heat) to those of greater probability (so of cold).

And...

> We may therefore identify the states of lesser probability with the past (i.e. with the Big Bang) and those of greater probability (via entropy) with the future.

Then in conclusion...

> The direction of time's passage (going forward) can thus be defined physically in terms of probability considerations (improving).

And lastly,

> The universe was (and is) passing from states of lesser probability, (again, the past) to that of maximum probability (so the future) via entropy, and the direction of this passage (is) defined as that (passage) of time.
>
> ~"The Evolution of Scientific Thought from Newton to Einstein" Aram D'Abro, 1927

Again as first noted herein, the most probable condition of my most probably disorganized (unattended) shelf number 1 is more likely. So, absent my outside work to organize (which causes the Opposite of Entropy), then my shelf number 2 will ultimately fall into disorganization and will also yield to entropy as currently demonstrated upon shelf number 1. Wow! I certainly hope I have convinced any and all the thermodynamicists and wave theorists (recall the possibly dissenting scientists) first mentioned!

And so to more detailed discussions of Entropy's specific current definition(s).

Given that heat to cold follows the same direction 'forward' as does the past through the present and to the future, then how and why do new suns and solar systems and galaxies (or even new elemental atomic structures) continue to be created?

Part Two, the "Practical Applications," follows. Please read on.

Chapter 20B

The Opposite Of Entropy, Part Two: Practical Applications

Back to shelf number 2 of my bookcase...

The very word "entropy" itself is absent from any dictionary in my (again, rudimentary) collection prior to the 1950s. <u>Webster's</u> and <u>Oxford</u> <u>University</u> <u>Press</u> seem to have published most of the early versions. But none prior to 1955 and that are upon my shelf number 2 contain specific definitions of "entropy."

And then, several do as follows re: Entropy:

> First proposed by Clausius (1865) in sense (as) 'transformational contents' of a system. Physics: The name given to one of the quantitative elements which determine the thermodynamic condition of a portion of matter.
>
> ~*Oxford University Press*,
> ed. 1955, page 619

Note: Prior to 1955, more specific definitions no doubt exist, but the above reflects an essentially unknowable concept of what entropy meant in 1955. Whatever were the "transformational contents" or were (or are?) the "conditions of a portion of matter" remain unknown and undefined. Then…

> A turning toward, etc. In physics; the theoretical measure of energy, as of steam, which cannot be transformed into mechanical work in a thermodynamic system.
> ~*Webster's New Twentieth Century Dictionary*
> Second edition, ed. 1957, page 608

Note also: What is possibly meant by a "theoretical measure" is unknown and undefined. However, both definitions implied three general requirements or essential conditions by which entropy could be defined and/or identified.

1. A recognizable, or better, a <u>definable system</u>, which was, early on, a thermodynamic, mechanical system. More specifically, Entropy was first recognized as a measurement of lost potential energies in <u>steam engines</u>.
2. A definable base of measurement with as near-perfect organization as possible wherein entropy could be regarded as zero, and then…
3. Subsequent base(s) of <u>dis</u>organization (originally, as measurements of lost heat energy) that could be compared to requirement 2.

Again, Entropy was first understood to have been a 'calculated' measure of 'lost' heat energy that did not produce a 'calculated' amount of work. This was first specifically measured when a calculated quantity of water was converted by a calculated amount of heat (into steam) in a contained and measurable device...again, a steam engine.

Such a thermodynamic system at rest understandably fulfilled essential condition 1 (above) but was also then, understandably, subject to heat conduction and friction, which were then essentially unmeasurable. So the early concept of entropy or of any measurement thereof remained relatively unclear and obscure...at least through 1957, and again per the dictionaries upon my shelf number 2.

Subsequently, since ±1958, *Webster's* in particular became both more specific, and at the same time, more expansive in its multiple definitions:

> Entropy... Physics: An index of—the degree in which the total energy of a thermodynamic system is uniformly distributed and—is thus unavailable for work.
> *~Webster's Dictionary of the English Language*
> 1992, page 326

Fair enough, but still confusing. Then *Webster's* continued with an expanded definition:

> In information theory, a measure of uncertainty of our knowledge.

Really? At that point, a measurable loss of heat-produced work was somehow related to a measurable loss of knowledge, perhaps even to a loss of intelligence? The exact meaning(s) of entropy had rapidly expanded toward the end of the twentieth century!

And then, into the twenty-first century, *Webster's* continued to both perfect its (older) definitions and, at the same time, add new ones:

> A thermodynamic measure of the energy <u>un</u>available for <u>useful</u> work in a <u>changing</u> system.
> ~*Webster's New World Dictionary*, fourth edition, 2003 (emphases added)

Note again: The "<u>un</u>availability" and "useful" (as opposed to <u>un</u>useful, perhaps?) work and "changing" system(s) were again most difficult to define for the classic meaning of *entropy*. But then *Webster's*, vintage 2003 again, went on to add that entropy's definition has been broadened further to include:

> A process of degeneration with uncertainly, chaos, etc., specifically when regarded as the final—stage of a <u>social</u> <u>system</u>. (emphases added)

Per 2003, *entropy*'s definition(s) included (1) the dissipation of heat energy, (2) the diffusion (so then the implied loss) of knowledge, and (3) the degradation of society(ies) that might end in chaos! Good for entropy itself...but bad for an understandable definition thereof!

It is obvious that over an interim 157+ years (from 1865+) and its first observation (by Clausius) and since, now more or less, one hundred and fifty years since its initial definition(s), the meanings (now plural) of the word *entropy* (still singular) have been expanded rather widely. *Entropy*, first observed as the physical <u>absence</u> of potential work in a thermodynamic system, now may also describe an <u>uncertainty</u> of knowledge and/or the <u>presence</u> of societal disorder that results in "chaos, etc." Of note here: A number of other, interim, and expanded concepts of entropy have been added since as well. Therefore...

As a science writer, it is easy to become confused. The concepts of entropy have morphed into all sorts of 'lost' or 'more diffused' but always 'less-organized' states of most any system(s)—whether thermodynamic or intellectual or societal—that are imaginable. Okay, then...

Entropy seems to apply when the organization of any system lessens and becomes <u>less</u> defined, or <u>less</u> hot, or even <u>less</u> understandable. Becoming <u>less</u> measurable and less organized is the key to most all current definitions. Entropy has become synonymous with increasing <u>disorganization</u> therefore. And if so, then what might be its antonym, "the opposite of entropy" (again, this chapter's title)? What causes or explains "anti-entropy," which would instead explain the <u>increased</u> organization of or within any system?

Anti (whatever) might express or imply any opposite meaning, such as 'anti-good' might imply 'bad' or that 'anti-right' might imply 'left.' Neither such definitions nor uses are conceptually logical. Notably, neither an acceptable word nor a proper expression of or for "anti-entropy" nor for "reverse entropy" is currently evident. Though entropy may indeed be reversed as the organization of any, previously entropic, system improves, there is no generally accepted term to describe such a process. Organization might be the result, but again, what has been or what is the process?

Regress for a moment and examine some original intentions and original definitions...these versus the more colloquial and current 'spoken' uses of several English words. "Bad" has recently been used to imply "good," however currently overused. There are and exist *carrots*, *carets*, *carats*, and *karats*. Whether a root vegetable (most enjoyed by rabbits) or an (obscure) proofreader's mark or a measure of a gemstone's weight or the percentage (within scale) purity of rare metals, the same, audible words apply… but with totally unrelated meanings. To "desert" means to "abandon," but "desert" also implies lots of sand. As opposed to a "dessert," which implies foods with lots of sugar…and "just desserts" that imply lots of consequences!

Accordingly now and as currently defined, what <u>does</u> *entropy* mean? Exactly?

And the same confusions extend also to the changing usages of the very same words in common conversation. "Discrimination" (the noun) meaning "to discriminate" (the verb) was originally "a positive trait and commendable behavior." It was <u>of</u> <u>benefit</u> to have been "discriminatory" (the adjective) in the past. Such behavior was synonymous with being "selective" (reference: *The Concise Dictionary of English Etymology*, Skeat, 1993). It is now <u>of</u> <u>detriment</u> and is <u>politically incorrect</u> to be socially discriminatory currently. Lots of similar examples exist.

English words, more specifically their meanings, do change over time…in particular as has "entropy." Now currently (in specific regard to physics), entropy is applied to the behaviors of atomic and subatomic particles, and so then to atoms and molecules, also compounds and solutions so upon the smaller scales. Entropy also and even more commonly applies to considerations of astrophysics and cosmology upon the larger scale as well.

Upon <u>any</u> scale, entropy has assumed a generally accepted meaning of increasing <u>disorder</u>. So then it is a <u>lack</u> of organization and then a <u>lack</u> of definition, which implies a <u>lack</u> of measurability and even a <u>lack</u> of understandability, all with corresponding loss(es) of heat energy, specifically of temperature(s) as implied. To be <u>lacking</u> (insofar as being less organized and/or colder) is the common requirement.

At this point, several more tangible and understandable examples best apply. A deck of (poker) playing cards represents an organizational system (requirement A), and its exact or original order (requirement B) may be obviously compared to its ultimate states of disorder (requirements C_1, C_2, C_3, etc.) as demonstrated:

First, witness an unopened deck...in its original box and in its original wrapper. It is both in suit and in exact chronological order (requirements A and B). Again, it displays zero entropy (for playing cards).

Second, entropy would be introduced as the deck was opened and cut once (requirement C_1).

Third, entropy would proceed and increase as the deck is shuffled (C_2, C_3, etc.).

Fourth, entropy would increase as the cards were dealt (C_4+).

Fifth, if the deck were simply scattered about upon the table or upon a floor. Or...

Sixth, if the deck were gathered up and discarded in the trash...then it might demonstrate total playing-card disorder = maximum entropy (C_5++).

As demonstrated above, it is easier to generally describe entropy conceptually than it is to specifically define it verbally or, if possible, to measure it mathematically. And...

The same deck might well be picked up (even out of the trash) and reorganized, so re-sorted manually (to discover any missing cards). Or, it may be re-suited and put back into numerical order via a successful hand of solitaire? It may then be returned to its original box, perfectly reorganized, thus reversing entropy back to requirements A and B. Why not? The various 'stata' of a deck of playing cards are easily manipulated. While...

In other instances, entropy, once achieved or once established, may be impossible to reverse. Two classic experimental examples come to mind:

Experiment number one, the first and irreversible:

a. Three identical beakers are placed in a row. The first (on the left) and the third (on the right) are half-filled with purified water. Beaker number two (in the center) remains empty. So all three are visually colorless, and there is a system (requirement A) established.

b. A permanent yellow dye is added to beaker number one (again, on the left), so its contents absorb all colors except yellow, which is reflected and is so observed. And then a blue dye is added to beaker number three (again, on the right) so its color blue is reflected and observed.

Note here: The three beakers then display yellow to clear to blue chromatically, left to right, which demonstrates the absence of any color entropy in the system. At this point, both requirements A and B have been satisfied. Then...

c. Beakers number one and three are both emptied into and so fill beaker number two, wherein its contents...

d. <u>Turn</u> <u>green</u>. While the exact chromatology and description of the resulting optics are complicated, the observed entropy of the resulting solution is not. And...

e. <u>No</u> amount of stirring nor any means of <u>un</u>mixing will ever cause to reflect the separate colors of just yellow or just blue light to reappear from beaker number two. Lastly...

f. The entropy (green) of the two original colors is established and fixed in beaker number two and will remain (as long as its solution does). Requirement C has been satisfied.

"So what?" may be asked. "Mixed fluids disperse and share their contents and characteristics in this fashion always." There is little or no consideration of lost heat. Whether the odor of onions sautéing in a kitchen or of a (bad!) cigar being smoked in a poker room/game or of diesel exhaust fumes most anywhere in Mexico City, the resultant pseudo-entropies are understood and then will become more unobservable over time. The onion, cigar, and diesel odors will dissipate in due time to become much more entropic so will then, ultimately, be undetectable. So do fluids, unrestricted and again in time, then actually diffuse and diminish and become less entropic thereby? Not so. The 'measurability' of entropy may diminish, but the diesel exhaust will never return back into buses, and the stink will never return to the (nonexistent) cigar. Nonreversibility is understood.

Witness Earth's atmosphere...also behaving as a fluid. It will require a great deal of time and substantial absorption and extensive photosynthetic processes to reduce its currently increased CO_2 pollution back to its normal concentration. Entropy of the gaseous contents of Earth's atmosphere is complicated and based upon a proper 'mix' of primarily nitrogen, oxygen, CO_2, and others (primarily "pollutants"). Its standard 'mix' of approximately 80 percent nitrogen, 18 percent oxygen, and 2 percent 'other' gaseous contents represent its requirement B. Any changes are, by definition, entropic...and unhealthy, and complicated!

A better example of entropic permanence would be better explained by an experiment number two below, which may be technically reversible, but deals instead with irreversibility and solids:

    a. A test tube is filled, first one-third full with white sand. Then another one-third is filled on top with black sand. So the tube's contents, as layered and as observed as <u>black</u> over <u>white</u>, are specifically organized (same as were beakers numbers one and three originally in experiment number 1b above). Then...

    b. The test tube and its contents are shaken vigorously, after which its contents then...

    c. <u>Turn gray</u>. And no amount of additional shaking nor any other external manipulation(s) will ever (?) in time reorganize the contents back into separate and distinct layers (nor into stripes nor into any other specific organizations) of just black and white...or will they?

However, there does exist the theoretical possibility, given enough time and enough shaking, some observable, specific organization of black over white might reappear, if only for an instant. The odds (better, the probability) of this occurrence must be similar to the scenario where enough monkeys typing on enough typewriters over enough time will produce a copy of the Magna Carta! Nonetheless...

The entropies evidenced and observed in the (forever green) center beaker and the (forever gray) shaken test tube are and will remain <u>fixed</u> and <u>unchangeable</u>. "So what?" may be asked again.

"So what then might want to or be able to describe reverse of disorganization or the reverse <u>of</u> entropy?" This is the next obvious question. And so to Part Three...

Chapter 20C

The Opposite Of Entropy,
Part Three: Contropy

"Evolution" has been seen as demonstrating the <u>reverse</u> of entropy. Physicist Martin Schrödinger, in particular, referenced that the process of evolution might appear to demonstrate "disentropy." It is obvious that as a species evolves, its structures, functions, and forms become <u>more, not less,</u> specific. All aspects of an evolving species tend to become <u>more</u> complex and <u>more</u> unique over time. Via these natural processes, species become <u>less</u> entropic by observation and per the general definition, therefore.

Any species' DNA, in particular, becomes <u>more</u> individualized and <u>more</u> characteristic, and so it displays <u>less,</u> or reversed entropy over time…as would be in agreement with Mr. Schrödinger's observation.

Counter to such an observation as above is the more general <u>perspective</u> that 'first-life' must have originated in one molecule and from essentially one place and at one time…presumably via said original molecule of primal DNA that was the progenitor of all subsequent DNA molecules since. So then life itself has, over the same time and now extant in essentially all earthly environments, expanded in so many and in such diverse ways and different directions, so expressing <u>increasing</u>, not decreasing, entropy. And further…

The original event, the original moment that was the start of life on Earth, would be highly <u>im</u>probable, much the same as was the Big Bang, and would and has expanded its existence, much like our universe has, and continues to expand its own. So…

Depending entirely upon the observer's specific viewpoint, then life, and in fact, any living system, may be considered as becoming <u>more</u> expansive and <u>more</u> diverse over time to express <u>more</u> entropy. While, and at the same time, life and/or any living system becomes <u>more</u> specific, <u>more</u> complex, and <u>more</u> unique, therefore <u>less</u> entropic simultaneously?

An obvious contradiction in perspective(s) exists.

Evolution: Specifically, its demonstrated entropic <u>function</u> is quite opposite to its <u>dis</u>entropic and more specific <u>form(s)</u>. As such, evolution of species presents two opposing entropic distinctions, and can be argued either way.

Similar observations may be made of the similar evolution of human societies and politics (reference Part 2, *Webster*'s expanded definition 4b in particular)…and of human languages most specifically. *Homo erectus* (first 'man' to stand) may have had and may have used any number of 'grunts' or 'snorts' as his or her early words. Imagine then or later attempting to explain "relativity" or "quantitative easing" even to a (vastly more advanced) Neanderthal! So words and therefore languages have expanded in their uses enormously as an example of social or 'linguistic' entropy. While, and at the same time…

Languages (and their corresponding speakers within their individual societies) have, again at the same time(s), become much more specific and much more complex. So here again, the processes of <u>entropy</u> and <u>disentropy</u> may be seen to exist simultaneously. Recall again currently how best to explain "quantitative easing"! And the contradiction persists.

Evolutionary processes must be, if only according to exact definitions thereof, confined to <u>living</u> systems…so then to plants and animals and their related developments and behaviors. Examples of entropy increasing or decreasing in any organic, or living, system therefore are dependent upon any observer's perspective(s).

An analogy may be made to the start of the simple "yes or no" guessing game, "Twenty Questions." The first rule is the requirement to first specify "animal" or "vegetable" or "mineral." Questions must apply <u>either</u> to organic/living systems and subjects or to inorganic subjects. Without such specificity, twenty questions are not enough.

Truly entropic processes, as observed to be specifically decreasing in organization (and so in temperature), to result in a relative state of greater disorganization and/or disarray, apparently need be restricted, again per exact definition, solely to the inorganic or "mineral" world. Accordingly…

The purpose of this Part Three of this chapter is to explain the restriction and containment, and then, most specifically, the reversal of entropy within <u>in</u>organic systems, or processes, again within the "mineral" world.

Inorganic 'stuff,' so then anything(s) 'mineral' and therefore of any mass, large or small and <u>when left alone</u>, so then when unaffected by any additions of the four currently understood energies, said 'stuff' is understood to lose temperature and so to fall into disorder and into disarray over time. Such again is the requirement of the Second Law of Thermodynamics and is the definition of entropy. However…

Why then, or better, <u>how</u> then do so many primal and very specific and more complex systems of organization(s) exist instead? Our universe is quite specifically organized on all 'scales.' All atoms and both their primary particles, so to include all elements and compounds, have had plenty of time (say, ±13.75 billion less ±380 thousand years) to disorganize themselves...but have not done so. It has been 'business as usual,' again for a net ±13.42 billion years, on the small and very small scales. Atomic, so elemental organizations exist quite specifically...and in very precise and increasingly complex (as elemental structures compound) ways. "Recombination" did not express an increase in entropy... rather the opposite. Atomic structures and photons, in particular, organized themselves, so their individual entropy was massively reduced.

Protons and neutrons, for examples, do, on the small scale, exist most specifically as tiny packets of three very specific quarks...much the same as do moons and planets and stars, as relatively larger packets of matter, exist as well-organized solar systems on the larger scales. One might observe the ultimate existence then of black holes, remnants of vast stellar collapse(s), as being the ultimate expression of ultimate reorganization on any scale. Certainly, the formation of black holes, to include such processes underway currently at the centers of most galaxies, do express disentropy within our dimension.

But, is the term "disentropy" the proper reference? Might the processes of entropy versus reverse entropy be compared to the processes of fission versus fusion, or of radiation versus absorption? Fission in reverse would not properly be referenced as "disfission" any more than would absorption be referenced as "disradiation." So with proper respect to Mr. Schrödinger, does "disentropy" make about as much sense as "disevolution" might make to Mr. Darwin?

The reverse (so the opposite) of entropy would seem better referenced as "contropy," would it not? Contrary to entropy would seem, most logically, to be contropy...which has been and will be referenced hereinafter.

Proper explanations of contropy would deal first with the regulation and containments (two) on the very small subatomic scale, then to containment of structural integrity on the atomic scale, and so lastly to the actual reversal of entropy upon the large, cosmic scales. Involved in and upon each level are the four, to date, extant force energies as understood...again:

1. The weak nuclear force is the most esoteric and, accordingly, seems to be least understood. Transmitted by theoretical (so indirectly observed) massive bosons (W^+, W^-, and Z^0 particles), the weak nuclear force regulates the process of radioactive decay...whether and at what speed and what sequence elemental radiation does or does not proceed. It was, theoretically, the first to depart from the combined strong nuclear and electromagnetic forces (numbers 2 and 3 below), and is seemingly unrelated to the force of gravity (number 4). Being the primary force of subatomic behavior(s), it controls, better it constrains, subatomic entropy insofar as, without it, in theory, all elemental structures would, in time, convert to energy. Accordingly, an absolute chaos of material structure(s) might exist in its absence. A sort of absolute, totally energetic entropy would prevail...as perhaps it did at the very moment of our creation?

2. The strong nuclear force demonstrates its two primary containments upon the atomic scale. It is

transmitted and so applied by massless subatomic "gluons" (as may have been first observed and confirmed at SLAC?). Gluons (so bosons, in a sense comparable to photons) have two very primary <u>containment</u> responsibilities:

i. Gluons keep the primary, elemental fermions, quarks, currently the smallest known bits of matter, in order and always in groups of three within protons and neutrons. So gluons at the atomic scale of 10^{-13} centimeters (one fenometer) retain the structures and so then retain the architecture of nucleons. Absent this containment, one might imagine a universal cloud of quarks just 'floating around' and in total disarray. This potential disorganization, not unlike the disorganization that might result from the absence of the weak-force bosons, would result again in universal chaos on the subatomic scale.

    Note here: The mechanics (and mathematics) herein above are complex. But both nuclear forces apply order and regulation (weak) and containment (strong) both of and within atomic and for atomic substructures. And without either one, any thought of any order at all within matter would be impossible. All atomic structures and substructures would seemingly come apart and disorganize without the presence of both nuclear forces. And...

ii. Gluons also contain the architecture of all atomic nuclei. Consisting of neutrons (with no charge) and of protons (each with single positive charges), electromagnetic repulsion between the protons would disintegrate all but hydrogen nuclei, absent the containment provided also by gluons. Again, absent such containment, atomic nuclear particles (nucleons) would come apart and might well exist as another totally disorganized cloud of same. Chaos on the small scale would result...but is denied, nonetheless.

And the same (or similar?) result may be argued for...

3. The electromagnetic force. Actually, it is the only force of the four currently reorganized that exhibits both attraction <u>and</u> <u>repulsion</u>. One might conceptually observe that the 'electro' function repulses and that the 'magnetic' function attracts. Its forces are transmitted and so demonstrated by massless photons. Given that opposite forces attract, magnetism keeps electrons (each carrying a singular -1 negative charge) in contact with and in proper orbits around atomic nuclei, most specifically around an equal number of protons (each carrying a singular +1 positive charge) therein. The number of electrons orbiting any non-ionized elemental atom are exactly equal to the number of protons within any nucleus thereof. Therefore, absent the electromagnetic force of attraction, electrons would detach and separate themselves (and from each other due to their repulsion also), again into a chaotic cloud of primary matter particles. Again, the implication herein and otherwise on the small atomic scale would again be to result in a state of near total entropy...resulting in no atomic structure(s) at all.

Thanks to the magnetic containment so provided, said structural entropy is denied.

Of summary notation then, the first three force energies above deal with the following:

1. First: The regulation and control of radiation insofar as to when elemental structures may or may not convert to energy. Such may be envisioned, perhaps, as the control of 'quarkian behavior' and is explained by quantum flavor dynamics (QFD), the technical terminology that explains processes that do or that do not result in radioactive decay. TBD. Such is the function of force number 1, the <u>weak</u> nuclear force.
2. Second: Containment both of quarks within nucleons <u>and</u> then the same of nucleons within atomic nuclei are the functions of force number 2, the <u>strong</u>-nuclear force. Said functions are described technically and mathematically per quantum electrodynamics (QED), which maintain both the internal structure of nucleons and the external architecture of atomic nuclei. The strong force therefore provides the basic architecture and organization of atomic nuclei...so then of the core structures of all matter.
3. Third: Containment of electrons in proper orbit(s) around atomic nuclei seems to be a primary function of force number 3, the electromagnetic force. Said containment still allows electrons to bind with other elements and so to share orbits in the formation of chemical compounds, which again demonstrate <u>contropy</u> in the process of their formation. No doubt, compounds can break down and diffuse, so demonstrating entropy. However, at the same time(s),

new and more complex molecular structures and compounds are created simultaneously. Thinking back to our universal creation, "inflation" may be regarded as the primary demonstration thereof? TBD. Regardless, absent the attraction between electrons and atomic nuclei, again, matter would break down into a cloud of totally entropic primary particles. The strong nuclear, not unlike the weak nuclear, forces provide order on the atomic and subatomic scales…via restriction, regulation, and containment. But notably…

The two nuclear and the third electrodynamic forces as above, by themselves, do little (or nothing?) to actually reverse entropy (i.e. to provide contropy) in or upon the large solar, nor upon the very large galactic, nor upon the very, very large universal scales. Such <u>contropic</u> functions on these larger scales are the responsibility (result?) of energy…

4.   Fourth: Gravity. Whether in fact technically the "weakest" of the other three force energies or not, gravity is certainly the most "omnipresent." If space may be defined as the absence of all matter (so of all 'un-dark' matter), neither nuclear forces nor the direct attractive nor repulsive effects of the electromagnetic force may be controlling nor even evident therein. But gravity most certainly is, and is evident everywhere…including, however weakly, on <u>all</u> scales. Given gravity's immediate effects are directly proportional to existing 'in-space' masses and then reversely proportional to the square of their distances of separation only, its effect and ultimate control is only diminished, never eliminated.[*] The

mutual gravitational attraction of almost totally entropic, interstellar gases and 'dust' result in the formation of quasi-stellar (planets) and stellar objects, which again, subject to gravity, may create densities and temperatures sufficient to ignite nuclear fusion. Note that gravity, per the functions above, <u>creates</u> heat, which is the opposite of entropy. Also in the formation and organization of solar systems, gravity provides, or better, <u>is</u> in effect, the centripetal force of same. And as may be proposed herein, so is gravity responsible for the formation and ultimate fate of galaxies and of other, possibly very large universal structures. And lastly, in supernova events, whereupon stars expand then contract and self-destruct to become either neutron stars or, ultimately, (still theoretical) black holes, none of these stellar events, whether observed or theorized only, would or could have occurred without and exclusively because of gravity.

    * "Elimination" of gravity may instead be evidenced as its possible transition into heat energy at a "similarity"…as discussed in the notations that follow.

Gravity itself is the only force of, and therefore is the reason and mechanism for the <u>reverse</u> of, and so for the <u>Opposite of Entropy</u>. It is again the only force on the larger cosmic scale that again can organize nearly entropic 'clouds' of elemental gases (primarily of hydrogen and of helium) and other 'clumps' of (poorly defined) interstellar 'dust' into stars. And stars are cosmic 'furnaces,' so the sources of all other heavier (lithium++) elements. <u>Gravity</u>, so operating to <u>increase</u> organization, complexity, and heat, serves as the force of <u>contropy</u>, which is again the opposite of entropy, which instead <u>decreases</u> organization, complexity, and heat.

Note also: Our universe must have originated from the <u>most</u> <u>specific,</u> so the <u>most</u> <u>improbable</u>, therefore least entropic of all structures...the Big Bang. No doubt and logically, some gravitational event involving a mass or masses of unimaginable size reduced to a point, a singularity, so then to a very, very, very <u>big crunch</u> (as observed from its other 'input' side), which was simultaneously expressed by the Big <u>Bang</u> (again as observed from our 'output' side). Regardless of any reader's perspective here, <u>all energies</u>, including most specifically <u>gravity</u>, had to have been initially present at and within and somehow prior to the Big Bang. That our universe has been expanding and cooling since expresses entropy generally, which is the loss (actually, is the dilution) of heat and explains the forward, unidirectional arrow of time. Granted. However...

Within, and most specifically during our universe's recombination (again, a seemingly misleading term... since there was one <u>com</u>bination only) and since via stellar (and then galactic) formation(s), <u>contropy</u> has since and is occurring simultaneously, but without affecting time's arrow. TBD.

Imagine only such a primal event and then such expansion and universal dilution in the absence of gravity. What a cold and chaotic place this, our universe, would be. Given the same ±13.75 billion years of expansion via inertia only and again, absent gravity, there would most likely be no large-scale organization(s), and surely would be no life at all. So, even to imagine such a scenario may be regarded as impossible.

Therefore, the three force energies and gravity as currently understood again apply:

1. restriction and order of and for atomic radiation (weak nuclear),
2. atomic nuclear containment (strong nuclear),
3. atomic architecture (electromagnetic), and
4. contropy…which must manage the speed of and provide for the reversal of entropy and cosmic organization(s) on the larger scales (gravity).

Absent any or all of the first three, any and all atomic structures would lack any architecture or organization(s) at all. Absent the fourth, gravity, any sense of 'stuff,' however otherwise organized or disorganized, would be and would remain very cold (assume -273°C), very dark, and totally chaotic.

Entropy would indeed prevail. But…

If, therefore, gravity can and does limit the speed of universal (so technically, of spatial) expansion, then it must also control the speed of entropy. And, if entropy defines and/or controls both the speed and direction of time, what then is the relationship between gravity and time? Dr. Einstein's expression of relativity deals primarily with light and time. Might then time have a different(?) or a deeper(?) relationship with gravity?? TBD.

In conclusion:

Note A: It remains of absolute importance and is (are) the basic requirement(s) for the absence of chaos and for the limitation of entropy that the nuclear (weak and strong) and electromagnetic force energies 'apply,' and for them to, in effect, "do their jobs." But they do not directly reverse entropy. So then they cannot create, nor can they directly affect contropy. Only gravity can do (and does) so. The primary effect of gravity must be the expression of contropy.

Note B: If the ultimate result, so then the ultimate expression of entropy is "chaos," then perhaps Steven Hawking's description of black holes is dimensionally correct? Per Dr. Hawking: If black holes are instead the ultimate expression of entropy (so not of contropy, as herein above presented), then a black hole at its singularity must have converted or torn asunder all matter that we understand at and within our world? Such destruction or reduction or conversion of matter must be directionally smaller or dimensionally 'downward.' (Refer to chapter 17, "Dimensional Proportionality.") So does the physical maximization of contropy (as evidenced by that point of infinite pressure and infinite temperature) result in dimensional entropy? Again, such would require dimensionally smaller universes that may exist upon the end(s) of Dr. Sagan's (or of anyone else's) finger(s)! How else might Dr. Hawking have explained the black hole entropy that he understood to exist?

Note C: To better understand the mechanics as above, what exactly must be focused and so possibly maximized at/within a singularity? Answer (by logic): Gravity is so required. And gravity is known and has been shown to be inversely proportional to/with temperature—which is heat. It is logical to assume, therefore, that gravity maximizes (gMax, perhaps) exactly at/in the singularities that must form exactly within the centers of any/all black holes.

Note D: To advance this logic then, the matter so focused and so compressed at this point must convert to energy (as is predicted, better is demanded by $m=E/c^2$). And such energy must first (and only?) be expressed by/as heat, which heat/energy is subsequently (effectively instantaneously?) again reexpressed and evidenced on the (again) "expressive side" of any singularity.

Note E: Assuming such energetic dynamics, the net gravity (presumably less than gMax) that remains would remain at or very near absolute zero since most or all the heat has been expelled or transferred, and said heat evidenced then as/at tMax would have to carry with/within itself both the weak and strong nuclear and electromagnetic forces into a new (or alternate) dimension.

Note F: Said "new" (or alternatively smaller) dimension would experience its very own subsequent Big Bang. TBD.

Note G: Specifically, Dr. Hawking did propose/conjecture that the temperature at/within black hole singularities is very cold, which would explain the absence of heat (again, the converted mass/matter) that must have "passed through."

Note H: With reasoning again that gravity is experienced and can exist only in the presence of mass (matter), then what must occur and/or what must remain at a singularity after any initial conversion ($m=E/c^2$)?

Note I: In summary, to better explain the process of contropy herein, consider what must have been focused and potentially if only momentarily maximized at/within any singularity. The answer again by logic: Gravity. And (maximum?) gravity is thought (known?) to be possible only in/with the absence of temperature, which, again, is the absence of heat. It is perfectly logical, therefore, that gravity maximizes (gMax again) to form singularities, which because of the resulting (or simultaneous?) supernoval explosions result in similarities that essentially expend or even consume gravitation into heat, which is (also simultaneously) expelled.

Note J: Again, this must occur only at/within black hole singularities wherein focused (maximized) gravity is boosted (therefore momentarily super compressed) by the inward (read: the super contropic impulse) force and reversal of the supernoval (read: entropic) outward forces of such an explosive event.

Note K: To advance the logic therefore and therein: The matter (so all the collapsing star "stuff") so focused also and super compressed also at that point (say at dMax) must and does convert/revert into energy (again as explained/expressed as in $E=mc^2$, which <u>reverts</u> or <u>converts</u> into…).

Note L: Heat, which is simultaneously expressed on (or "in") the "other" (so <u>white</u>) side of any <u>black</u> hole singularity. And if such mechanics do exist, then…

Note M: The transformation of gravity itself into heat, which explains/describes an energetic "similarity," creates an effective <u>temperature vacuum</u>. Recall:

- Heat is <u>not</u> the absence of cold.
- Cold <u>is</u> the absence of heat (and again is the vacuum).

So then:

- Contropy is <u>not</u> the absence of entropy, but then,
- Entropy <u>is</u> the absence of contropy (so another "vacuum" of sorts).

Accordingly:

Note N: Contropy and coldness (being the opposites) are or must (however possibly and/or mechanically/physically) be balanced and equated. And this sounds very much the same as a "super conservation" requirement that neither (new) matter nor (new) energy can be either created or destroyed. And therefore that:

- Matter is converted to energy at a <u>singularity</u> and <u>in</u> and <u>at</u> that point...
- Gravitation is converted also into heat at that corresponding <u>similarity</u>, which heat is then...
- Expressed as a Big Bang on/at the "white" side of any "black" hole, which cools over time to result first in...
- The reorganization of said heat energy back into the other three actual force energies and subsequently back into...
- Matter, so $m=E/c^2$, which is (so was) demonstrated as "baryogenesis" in our infantile/initial universe. And which matter...
- Reconfirms or recreates and demands gravity, which is thereupon...
- Omnipresent...until refocused.

Note O: Even Stephen Hawking did (again and prior to his death) propose several elements, or better, several conditions that must exist in/within black hole singularities that:

1. The temperature(s) within such hole(s), assuming their loss of heat, would have to be very cold (and vacuous)...possibly at or very near absolute zero. And that...
2. Over time and with the accumulation of additional (read: new) matter, said hole(s) would gradually...
3. Infill (or themselves effectively "evaporate") as they warm, and confirm that...
4. Black holes form from/inside/at the exact centers of supernoval explosions, and (that) they first lose

and then regain their heat energy in this process. So then…

5. They (black holes) must "warm up" and so "evaporate" over subsequent time(s) to simply cease to exist.

Note P: So however "best" or most "accurately" is explained, the most evident and primal form of all energy(ies) is heat, which must build up via contropy at and within black hole singularities, and then must disperse or cool via entropy on the "other sides" or possibly "other dimensions."

Note Q: Exactly how the balance of gravitation must go/ diffuse from its maximums to a minimum while maintaining its natural distribution at/via any corresponding similarity is and must be a process that <u>will</u> require future explanation(s). This is one: there will have to be others for sure…perhaps in *Fizzicks 201* or elsewhere.

Note R: Contropy is obviously in process in our known universe and via these process(es) must be creating smaller dimensions, which result in higher densities and <u>create their own heat</u> as a result.

Note S: Such "progressive" scientific thought is exhausting…as perhaps this author and his (so "my") readers may be.

Further discussions, again in *Fizzicks 201*, seems sensible…yes?

Chapter 20D

Being In The Same Place
At The Same Time

Unique then to contropy, one may propose that it can cause and/or result in gMax (maximum gravity) and tMax (maximum heat energy) and pMax (maximum pressure) and dMax (maximum density)—all to occur/exist at the same time and in the same place, which "cannot happen" and "is not permitted." TMax and gMax cannot coexist.

So...

They don't.

If not relocated and reexpressed then within or at another dimension while in the same <u>locations</u> (i.e., different spaces within spaces) but again in the same original dimensional place (i.e., a singularity), where else can they/either one "go"?

TBD...which may require a large expansion of this Chapter 20D indeed!

Chapter 21

"Pi"...Is An Irrational Number

Pi is perhaps the most famous and widely used mathematical ratio in existence.

Pi is also an "irrational" number...and is (in fact as expressed) also the fifteenth letter in the Greek alphabet...π.

Pi's public 'history' is well known, but its origin, perhaps, is not...not so well understood.

Archimedes (the Greek mathematician along with Euclid and Pythagoras et al.) first publicly computed pi in or on about 250 BC. His intent was to find the ratio between the diameter (think width) of any circle and its corresponding circumference. Simply...

Make a line, or take or find a rod or even a straight stick, so almost any 'linear' kind of thing with a fixed length. Measure it. This becomes a "diameter" when...

It is spun exactly from its center upon a flat, two-dimensional surface, and its end points scribe out a closed arc, which is a circle. Measure it. This then becomes a corresponding "circumference."

Note also: Any line or rod or stick of fixed length (so any diameter as above) may be spun in all three aspects, rather than in two only dimensions, so in all directions, while keeping its center at and upon a fixed point. This is more difficult to do, but results in a <u>sphere</u>. And spheres exist as the basic and primary three-dimensional structures in our universe. Note then: Any 'equator' of any same sphere will measure with the same circumference(s).

Spheres are simply spun circles.

Archimedes recognized this and found that he could divide any circumference, however created, by its corresponding diameter...so its width...and that the answer was always the same: 3.14159...n, which is pi. Always. It never changes. Further...

The computation of pi never ends, hence the notation "...n." Its ultimate (never exact) answer, which has been calculated to thousands of decimal places, is elusive, and its calculation goes on and on, ostensibly forever. And because it does so, computes continually and never exactly, pi, the answer, is considered "irrational." It is so considered also because no fraction exists that expresses this exact ratio... exactly. Other irrational numbers (inexact calculations) exist as well...like the square roots of two and of three, for examples. While...

On the other hand, the square root of one is one, and the square root of four is two...and both answers are understandable and exact. So then, both one's and four's square roots are considered "rational"...as are nine's and sixteen's and twenty-five's, etc.

One might inquire: What about the ratio: 1/3? Note that its <u>arithmetic</u> proof is: 1/3 + 1/3 + 1/3 = 3/3 = 1. But its <u>mathematic</u> proof is: 1 divided by 3 = .333...n. And .333 + .333 + .333 = .999 – which is not equal to one. Therefore, 1/3's <u>arithmetic</u> does not agree exactly with its <u>mathematics</u>. And, one may note that this disagreement (a mathematical contradiction of sorts) exists for any and all multiples of three-based fractions as well.

Note that: $1/6 + 1/6 + 1/6 + 1/6 + 1/6 + 1/6 = 6/6 = 1$.
But that: 1 divided by $6 = .16666...n \times 6 = .99996...$not 1.
And that: $1/9 \times 9 = 9/9 = 1$.
But that: 1 divided by $9 = .11111 \times 9 = .99999...$not 1.

And this contradiction continues for:

$1/12 = .08333 \times 12 = .99996...$not 1.
$1/15 = .06666 \times 15 = .99999...$and so on.

Therefore, such fractions whose mathematics integers repeat, but do not (and never will) compute to <u>one exactly</u> would best be said to produce "quasi-rational" answers...in this author's opinion. As compared to:

Totally "rational" fractions that compute rational answers and that do equate (exactly) to one, such as:

$1/2 + 1/2 = 2/2 = 1$, and so $1/2 = .5 \times 2 = 1$ or
$1/4 + 1/4 + 1/4 + 1/4 = 4/4 = 1$, and $1/4 = .25 \times 4 = 1$, same as
$1/5 = .20 \times 5 = 1$ and
$1/10 = .1 \times 10 = 1...$and so on.

However considered, pi at $3.14159...n$ is fixed and is always the same ratio, but at the same time, it is a never-ending calculation, which is inexact and is considered "irrational" therefore. Furthermore...

This ratio is always referenced to be pi, or π, which is not even a number! It is, again, the fifteenth letter in the Greek alphabet, and as such, is not exactly "rational" for that reason either...is it?

If confusing, also consider c, the <u>third</u> letter in the Arabic (our) alphabet. Also a constant, c stands for the exact speed of light. So in this sense, both pi (π) and c must be understood to

both represent very rational ratio(s) and/or measurement(s)…
as stated.

Fair enough. Now consider pi's simplest approximation,
which is 3.14, which can be roughly calculated, so reproduced,
by dividing 22 by 7. So, one might assume that 22/7 is close
enough to Pi's fractional value…but it is not. It is even less
exact. Note again that pi = 3.141592, while 22/7 = 3.142857.
If/when calculating the exact circumference of a bushing
necessary to house bearings for a two-inch-diameter drive
shaft, only pi works. If using 22/7, the bushing will be much
too loose!

Note also that 3.14 is also a date…that of March 14. Such
date is 'celebrated' by some (mostly math nerds) each year
as "Pi Day." Since our earthly dates are actually calculated
fractions, but this 'kind' of time is calculated as fractions of
the time it takes our planet to rotate and to orbit our star…our
Sun. But, not unlike the 22/7 calculation's inaccuracy, our
times and our dates are not exact either. Earth's orbit winds
up to be six hours short each year, so a new date, a new day,
February 29, must be added every four years. Hopefully,
Earth's orbital speed will not change!

One might 'recompute' this to say: "Let's just add six
hours per year and we're good…yes?" No, we're not good.
A six-hour-longer year would require about 3.9 seconds to be
added to each day. And, such longer days would not coincide
with Earth's rotation. In just four years, day would become
night…per our clocks. Plus…

The Earth's rotation is, due to the 'drag' created by the
friction and momentum of its ocean's tides, said to be slowing
very gradually anyway. The assumption might be then that
the 3.9-second-longer days will be evident…in time anyway?
The Earth's rotation, therefore, might one day coincide with
its six-hour-longer orbit anyway. It's complicated for sure…

but, our current measurements of our time(s) and dates are best left 'as they are'…not unlike pi in that sense!

And given there exist billions of suns much like ours (and billions unlike ours), plus multiple billions of planets (again, some being very much like ours), if they, our similar planets, are occupied somewhat like ours, they would have to have their own assigned measures of their times and their own dates—which obviously would be much different from our own. Therefore, considering the possibility of extraterrestrial intelligence(s)…

All their dates and times, like ours, would be necessarily arbitrary…so arbitrated by their planets' spin rotations and orbits. It is a sure bet that no other occupied planet's "geophysics" would be the same as ours…so their measurements and/or calculations of time would also be very different. But…

Their calculation(s) of the ratio of any diameter to its circumference would be exactly the same value as ours, which is again 3.14159…n. And this is true whether upon their planet(s) or upon our own.

We may consider our second to be to our 'gold standard' of time measurement. Note: There exist all sorts of our measurements of distances (feet versus meters versus fathoms and leagues, etc.), and of weights (ounces and pounds versus grams and kilograms and carats and karats, etc.), and of volumes and temperatures and of most any other sorts of things or of qualities or of quantities that can be measured. But…

Not for time. Seconds (and so milliseconds) are our sole and standard measurement of our time. All others (minutes, hours, years, and centuries, etc.) are merely expressed as in multiples of seconds. What then, exactly, is a second? How 'long' is it, exactly? What's its duration?

An (earthly) second itself is currently measured as being exactly equal to the time (so the duration) required for a cesium atom to vibrate 9,192,631,770 times. Really? Yes, really. Perhaps we can adjust in the future when either Earth's spin rotation slows or its orbit degrades or accelerates…however, not likely. There might be found an elemental atom or a particle that vibrates 9,192,700,000 times, for example…and we might be able to (or be required to?) adjust. But we won't. Cesium's 'count' remains our time's standard therefore. Seconds are (and must remain) seconds…our own, standard measure…always, in this author's opinion.

Conclusion: Most any behaviors, including any measurements, be they of time (again, seconds) or weights or distances and dimensions or temperatures or whatever… especially distances, such as sides and heights of pyramids (as below), are totally arbitrary. And if so, they cannot be just 'rational' but must be either situational or even coincidental, therefore. But, not so for pi, which is universal.

Of further note: The Great Pyramid of Giza has been subjected (and so measured at/per the centers of its mass) to very exact dimensional detail. It is exactly 146,515.174 millimeters high at its apex/top. And its four (not three) bases (sides) measure 921,452.72 millimeters in total perimeter… also exactly. Note then that 921,453 divided by 146,515 (both measurements above) equals 6.2891. Note lastly that twice pi equals 6.2832. Is this, the same calculation, correct to the 1/100ths, a coincidence?

Since the radius of a circle or that of a sphere is one-half its/any diameter, twice pi is the correct ratio of any circle's (or of any sphere's) radius to its circumference…as is expressed almost exactly by the Great Pyramid's design. So no, its measurements cannot have been coincidental! They had to have been intentional.

As first noted (in paragraph 4 of this chapter): Archimedes first (allegedly) calculated pi in 250 BC and in Greece. The Egyptians (Cheop's 'engineers') built the Great Pyramid in/around 2570 BC, obviously in Egypt…but ±2,320 years earlier! It, the Great Pyramid, is essentially, and so is both mathematically and architecturally the top half of a square-based sphere. Its dimensions and mathematics so confirm its design and its structure…almost exactly…and as it was obviously intended. So…

How did the Egyptians know and so calculate this ratio (again, twice pi) so exactly and so far (roughly 2,320 years) in advance? What did they (and how did they) know then that which Archimedes 'rediscovered' ±2,320 years later? Ergo…

Archimedes must not have been the original person or entity to have calculated the most widely used (and most universally accepted) ratio in existence. Recall again, any and all 'other mathematicians' that either do exist or that have existed upon any other similarly 'habitable' planets are now and were then subject to this same circumference/diameter ratio, which again is (our) pi = (everyone's) 3.141592…n. In other words, pi to us is whatever (symbol) might be to 'them'…but it is always the same set of numbers, however stated. And these 'other' mathematicians and scientists must also know or have known what c (again, lightspeed) is to us as well. However 'they' quantify either of these constants and, regardless of the mathematics involved, the answers must be the same…universal constants, and including twice pi, or 6.283184…n as well. So…might the Egyptians have received some 'outside' help…or not?

Words—and so news—of universal constants seem to get around quite easily! How 'wise' (so, how rational) must the Egyptians have been otherwise some 2,300+ years ahead of the Greeks! Again, neither the Egyptians' knowledge nor the Greeks' calculations were by coincidence…and it is not likely either that they were the result of an 'alien visitation'… as theorized by some. Or were they?

Go figure! Who <u>was</u> first to know? And how?

Back to the most simplified quantification of pi, which is 3.14. Note again that 3.14 is also an earthly date: March 14th. March 14th of any year is again celebrated by some (mostly math nerds) as "Pi Day." March 14th also has to be a lot of folks' birthdays…and one in 1879 (a particularly arbitrary year) was (is) notable for sure. Specifically, March 14, 1879, was (and is)…

<u>Albert Einstein's</u> birthday…and Dr. Einstein is perhaps <u>the</u> most rational man born to date…again, on 3.14. Coincidentally?

All the above have been presented as a mix of facts and histories and calculations and conjectures and one apparent coincidence (that of March 14, 1879). So what then, if anything in particular, may be deduced? Rationality <u>is</u> what it <u>is</u>. It must be. That is <u>rational</u>. Therefore, anything rational that results in or creates anything also rational itself makes <u>extra</u> good sense.

Rationality × rationality = rationality^2…by deduction.

Conversely, then, anything <u>irrational</u> by itself that results or creates anything also <u>irrational</u> also must make <u>some</u> sense…even if a bit less sense. Equating this logic directly to any 'double negative' and indirectly to the logic that: "The enemy of my enemy is my friend," then…

Irrationality × irrationality = rationality…also.

How else did the Egyptians know how to design their pyramids?

How else might Albert Einstein <u>not</u> have been born on 3.14 of any year?

Are not <u>all</u> circles' circumferences so exactly 3.1416…n times their diameters?

Why would (so how can) photons (so light) proceed at an absolutely constant velocity…and how can they continue to do so without any influence of time? Or of distance?

Furthermore, and for a slightly less than rational consideration: The digits of 1879 (again Dr. Einstein's birth year) total 25. The square root of 25 is exactly 5, which is a very rational number (so one-half of our 10-based system of mathematics)…and a very coincidentally rational result again for the good doctor.

Dr. Einstein's birth date's mathematics above both express their same kind of rationalities…and do so both via coincidence and according to computation. Note also that the computation of a square root is really another form of conceptual division. So, 1879 becomes both relative (and irrationally rational) in that sense…perhaps?

That may be what "relativity" is all about…yes? More about 'observations' first, but said observations must be confirmed by very specific mathematics, just the same as the exact measurements of the Great Pyramid have confirmed.

Were he here to reply, no doubt Dr. Einstein would so confirm!

Relativity, however vague, must have to be rational too… as is a <u>fixed ratio</u> of any <u>diameter</u> to its circle's <u>circumference</u>. But pi is itself and remains an 'irrational number,' nonetheless. It can be confusing, therefore…

Again, go figure. Most everyone else already has!

Every circumference of any circle and every equator of any sphere in our universe is and always will be 3.141592…n times its diameter. End of story! And it's nice to know that '<u>everyone</u>' agrees!

Chapter 22

Dear Ian…"Life Force"

March 23, 2021
Dear Ian,

Thanks again for your reference to Rupert Sheldrake's book *Morphic Resonance*. I have now read it twice…but it is still without my complete understanding. "Formative causation" is the basis of his writing and is the primary explanation of his thesis. Formative causation is difficult to define and more complicated to explain, however.

Rupert's son, Merlin, has since written *Entangled Life*…a copy of which I recently sent to you. Merlin's thesis, particularly upon the function(s) of fungal mycelia (more simply, upon mushrooms' roots), is not really that different from his father's, in retrospect. To explain…

How do more primitive living organisms, plants in particular, 'know' how to do what they do? More simply, how do they do what they do with essentially the same exact repetition and with the same results every time? How/why do all maple leaves always assume their same classic shape, each and every year and with essentially no mistakes? In fact, the specific leaves of all deciduous trees are recreated identically every year. Their 'knowing' how to do this must demand some kind of very primitive 'thinking.' And thinking demands some form of intelligence. So to complete this reasoning...

Intelligence must, and does, require some form of life. And the progression of this intended logic makes sense:

- Repetition must involve some kind of knowing, and
- Knowing must involve some form of thinking, given
- Thinking must define intelligence, and understandably, then
- Intelligence requires life. And...

Life is repetitions. However...

This logic may be challenged by the original game of "Twenty Questions." One person (player number one) must think of anything physical, be it a substance or an object, and the other person or persons (players number two and up) must name what that substance or object is by asking player number one no more than twenty 'yes' or 'no' questions.

This game would be much more difficult to win if player number one did not first have to specifically state that his 'thing' was either 'animal' or 'vegetable' or 'mineral.' First then, consider the 'mineral' possibilities.

Minerals, like elements or just 'rocks,' are most likely to be explained by physics and/or inorganic chemistry and/or geology. But they also may be explained via a very abstract notion of a sort of informational capacity that may be explained by somehow (their) 'knowing' and possibly 'thinking.' Rocks, for example, and elements are inanimate... so being without brains, for sure. However...

Given silicon and oxygen and heat, they do (always, when in/under the same circumstances) produce silicon dioxide (SiO_2) or quartz. And when quartz is fractured and granulated, the result is sand. And a lot of sand exists... almost all being grains of quartz with essentially identical, crystalline structure and cubical architecture and size. While this may seem all too obvious...

Note also that sodium and chlorine behave much the same (as silicon and oxygen do), but they produce sodium chloride (NaCl) or salt. And they may do so without needing the same conditions but do so with much the same crystalline structure and cubical architecture and consistent size...as sand. How then do they, very dissimilar elements, 'know' how to do this? Being always consistent in structure and in size...but also always sand or always salt?

Another example involves carbon and (again) oxygen. When they combine, again, given their particular circumstances, they produce either carbon monoxide (CO) or carbon dioxide (CO_2). Both are colorless and odorless gases, but one (CO) kills oxygen-dependent life forms, while the other (CO_2) supports plants in particular. In fact, oxygen by itself, and in the absence of CO_2, can kill plants (the 'vegetables') and nourishes most all the rest (the 'animals'). And 'us'!

Perhaps this too may be considered obvious. But, all these 'results,' or inanimate 'behaviors' (possibly?) must imply some kind of original organization and intelligence. We only have to ask: "Who or what 'knew' how to do this, or how to orchestrate these processes in the first place?" and "Why?" All animate life forms since have had to adapt. How were the original 'rules' set? And by what method and by whom?

Okay. Given that the foregoing just above describe inorganic (so mineral) processes (so inorganic chemistry), then the other two options in "Twenty Questions" must be (by rule) either 'animal' or 'vegetable.' And these options involve the '<u>org</u>anic.' Therefore…

<u>Organic</u> has become almost synonymous with any chemical compound containing carbon. Are inorganic compounds, therefore, by definition, without carbon? No, they are not, necessarily. Organic processes (reactions, repetition, reproductions, etc.) all imply <u>living</u> processes. Inorganic processes remain to explain all <u>non</u>living reactions. This implies that carbon is synonymous with life…but is it really? Only? What instead may involve some method of 'knowing' or original 'thinking' in any organic processes? Something must explain the intentional repetitions that plants and animals display consistently. Rocks cannot and do not shed any parts and then grow new ones in any year. Plants do so each year…routinely.

Okay. I'm at the point of possibly confusing myself! Instead, I'd propose an experiment for the three of us (to include my wife, Sharon), but at your place, next year, on your property there in Idaho. You and I and Sharon will each need a clean paper grocery bag. We'll need to embark on a hike just to Challis Creek in front of your house, next July 15th, and each of us will have a specific instruction:

1. You will pick about a half bag of live aspen leaves… only, and right off the trees.
2. I will pick a half bag of alder leaves…ditto.
3. Sharon will pick a half bag of willow leaves.

Fair enough? Then we'll return to your kitchen with our bags kept quite separated. To proceed…

We'll need three clean and separate blenders. We can bring one or two. Into one go several handfuls of your aspen leaves. Into another goes the same quantity of my alder leaves. And into the third go Sharon's willow leaves. We'll then add approximately two cups of either your tap water or about half a bottle each of "Arrowhead" pure spring water. Either way, our leaves will be hydrated with exactly the same water. Only the particular species of leaves within, again all from your streamside, are different. And then…

We'll turn on the blenders—*et viola*—we wind up with three leaf soups or light-green blends of liquid: (1) aspen, (2) alder, and (3) willow. If at that time we had the laboratory and technical skills, we might find ever-so-tiny differences in our 'blends.' Your aspen soup might have a bit more sulfur, for example, while my alder soup might reveal a bit more iron, and Sharon's willow soup might contain a bit more selenium…all guesses, for example(s). Otherwise…

All our 'soups' will have otherwise essentially the same identical organic chemistry. Then we'll proceed to your patio/porch and, given a nice sunny week in July, spread our soups out on three identical glass panels to dry. And let's assume this takes ±3 days (and no rain nor wind nor contaminants, etc.). Then…

We'll scrape the three light-green and now dry crystalline residues off and into three separate containers—say, three identical bottles—each itself identified again as either (1) aspen, (2) alder, or (3) willow. Next…

We take our bottles to a chemistry (?) or to a biology (?) lab at either University of Idaho (your state) or to the University of New Mexico (in ours) or, best, to both…with the simple instruction(s) to: "Make leaves."

Clearly, given rehydration (with distilled water in both/either labs) and sunshine and clean air (now it's August), all the necessary ingredients and conditions should be available. Even new 'dirt,' if so required! All the chemistry and elements in the exact and correct combinations and proportions are there, at hand and in our bottles. But what is/are missing?

1. First is <u>life</u>. What we've sent to the labs <u>is</u> now without it and is unliving, so it is <u>dead</u>. And...
2. The second is <u>information</u>, which is the instruction, so then the 'knowing' of how not only to make leaves but also how to make the following:
   a. either aspen (only) leaves.
   b. or alder (only) leaves.
   c. or willow (only) leaves.

The sorry facts of the matter are:

A. Although all the ingredients are there and all the chemistry is provided, the 'life' is not. Had we sent aspen and alder and willow <u>seeds</u>, perhaps then? Why not? 'Vegetables,' so plants, specifically aspen or alder or willow <u>trees,</u> do this all the time, each spring. Their seeds have the 'life.' Their blended leaves do not. What has been lost in the 'blends'? And...
B. "Make leaves" seems simple enough, but not possible, before making 'trees' first. Again, aspens', alders', and willows' seeds do this with ease—prior 'knowledge'? Given 'life,' how would they (our three crystalline substances) know even how to begin? They don't have the information with which to begin. Our bottles' contents, sent to either lab, are both:
   i. dead, and
   ii. dumb...therefore...

There's no chance to "make leaves" out of our crystallized 'soups.' But again, <u>trees do this every spring</u> out of the same chemistry.

C. Assume the impossible (sorry if an "impossible assumption" is difficult to make). Let's say first, that life might be somehow reinjected into our crystalline soups, <u>and</u> that the instructions (so specific information) were available to "make leaves." How would each and every (so 'repetitive') aspen leaf be made...as distinctly different from every alder and from every willow leaf...be made? What is the necessary information to do this? Both Messrs Sheldrake, Rupert and Merlin, had and have their similar explanations, but could not and did not address the 'life' question directly.

Rupert's, again, is "causative formation," which reveals itself as "Morphic Resonance"—the title of his book. Merlin's is more about a nebulous but <u>living</u> information network, displayed by fungi and transmitted by their 'mycelia.' But both speak to the very hard to define:

1. Exact <u>repetition</u> that must involve some sort of <u>instruction(s)</u>.
2. Said <u>instructions</u> must result in some form of <u>knowing. And</u>...
3. Knowing must involve some sort of <u>thinking</u>, if organic, or just of prior <u>instruction</u>, if <u>in</u>organic.
4. <u>Thinking</u> must involve some sort of <u>intelligence</u>, which can be processed organically in/with some kind of <u>brain</u>. And therefore...
5. If <u>intelligence</u> demands some form of <u>life</u>, then...

6. Life <u>forms</u> must have some form(s) of <u>brains</u> (however primitive), because…
7. Plants (so 'vegetables') <u>and</u> 'animals' can and do think (read: have the ability to process information).

Okay. Time out. This all may seem to be too confusing to be logical. And there exist only four known universal energies currently that can and have been identified as the following:

I. <u>Nuclear strong</u> force (keeps atomic and subatomic structures in order and in place)
II. <u>Nuclear weak</u> force (regulates how particles in [and within] atomic structures must behave)
III. <u>Gravity</u> (which demonstrates, if not explains exactly, how all masses attract one another), and lastly…
IV. <u>Electromagnetic</u> (which explains "all the rest"… actually, <u>almost</u> 'all' the rest).

Accordingly, "all the rest" does not include why our soups can or cannot make leaves. And, for certain, why leaves of aspens and alders and willows are made to be distinctly different, yet always each the same, and every time. Neither forces I through IV above can nor are they responsible for either these <u>abilities</u> nor the <u>decisions</u> (so the 'same differences'). So…

What is? What then is, so what then must be, if only by default, the force of life? How can there <u>not</u> be a separate and distinct "life force," therefore? Where is our energy number V?

Who or what designed (the <u>in</u>organics) or instructed (the <u>organics</u>) of the world, or better, if just within our universe. Who 'knew' how to operate (the <u>in</u>organics again) and how to behave (the <u>organics</u>) in the first place?

Is there a nuclear 'weaker than weak' force responsible that we currently do not understand? <u>Doubtful</u>.

Is there a fifth 'life force'? Unknown to date…but <u>probable</u>!

Again, science fiction may provide a clue. What was (is?) Yoda's and any Jedi's (and Darth Vader's) 'Force' that apparently enabled them? That force, whether fictional or not, may be very close to what might have been necessary for our universities to be able to "make leaves"…yes?

Why not? Stay tuned. George Lucas may pass away. But Obi-Wan Kenobi and Luke Skywalker will not…not ever. Okay? And to hell with Darth Vader! <u>Mis</u>uses of this force are not permitted!

So to summarize: Most of the open questions above that ask 'why' and 'how' only aspens and <u>not</u> alders and <u>not</u> willows, etc., may be answered 'by the genetics.' Genetics obviously involves genes. Genes are collectively made up of very specific 'strands' of RNA and DNA molecules, our DNA being organized with twenty-three 'pairings' that express any/all living characteristics in humans. A lesser number of pairings exist for less complex life forms…like aspens. Genes are themselves organized by three 'parts' that perform three critical functions:

EXONS	1. Exons, which specify <u>which</u> amino acid (so protein) sequence is involved. And…
INTRONS	2. Introns, which determine <u>where</u> such proteins are made. And…
??s	3. (unnamed) but 'regulatory elements' that have to determine <u>when</u> a specific protein being is made. And this, the 'when' of any process, may be the most mysterious. How might genes 'tell time'?

So, the 'which' (or the 'what') and the 'where' and the 'when' information is embedded in the (so within our) genes. And the genes—complex chemical molecules, therefore—are located in our cells' nuclei in structures referenced as 'chromosomes.' However...

Genes are still chemistry. They are uniquely organized molecules of relatively few elements...nothing particularly exotic or fancy...just uniquely different from other molecules insofar as how they apparently do know how and where and when to do what they do?

Genes can split up or divide and then recreate exact or almost exact copies of themselves. So, wait a minute! Why can't a molecule of water, H_2O, do this? Simple answer: Water is not alive. Genes are alive. So, it cannot be just the chemistry. It is by simple deduction, the existence or nonexistence of life. And as discussed above, carbon may or may not be involved or even present. For example, CO_2, carbon dioxide, is not alive and cannot—the same as water cannot—simply replicate itself.

Replication or reproduction is the one very obvious function that a living 'thing' can do and that a nonliving 'thing' cannot. And, living 'things' must begin with genes or cells that are alive and, by definition, organic. So again, how are they different from an inorganic thing that is dead? What is the difference between being alive (re: life) and not being alive (re: death)? Notably, it must be the same distinction between organic and inorganic substances with which to begin...yes? Inorganic things don't have genes.

So two very relevant and still unanswered questions persist:

1. What is life? What is the process whereby a living substance or living cell or gene is able to reproduce itself? (My question.)

2. How does it (or how do they), if and when living, preserve and replicate themselves exactly? (The Sheldrakes' questions.)

Question 1 is and should be simple: You and I are sitting at a table. One of us suffers a massive heart failure and dies and slumps in his or her chair…and is dead with no heartbeat and no brain activity being evident. So what then has been 'lost'? Conversely, the other one of us is quite alive. What does this alive one of us possess that the other dead one does not? Is our 'living' chemistry still <u>org</u>anic, while the other's is not? Is the deceased's chemistry simply not relevant or all of a sudden has it become <u>in</u>organic?

Question 2: If organic and alive, when a single sperm (male) cell unites with a single egg (female) cell, they, as if by magic, begin to divide and to replicate or reproduce themselves. So when and how does the 'offspring' of one of these original two embryonic cells simply 'decide' to become a toenail or an eyeball? However crudely stated, it must be upon the <u>instructions</u> of the gene(s) again. But they, genes, are again simply chemicals. Who or what is the <u>instructor</u>? Who or what is the <u>conductor</u>?

Chemical compounds cannot simply 'think' by themselves… can they?

This kind of inquiry appears to be endless. My time and paper and pen(s) and your attention are not! So, when we can understand how to make—actually, to <u>re</u>make—the leaves, only then can Messrs. Sheldrake explain how then we get either aspens or alders or willows, or somethings 'other'! And when (and if) our blends do (begin to produce leaves)…

Let's mix all our remaining leaf soups together and see what results! "Aspalderwills," perhaps?

Confusingly (but intellectually) yours, J. B.

Conclusion

Written in 2015. Revised in 2023.

To whom it may (hopefully) concern:

I've had seventy-four-plus years now to think about this: "What are we, *Homo sapiens*, really doing here?"

Light travels very fast, at a constant speed of ±186,282 (thousand) miles per second in space. And, contrary to logic, that is <u>always</u> light's speed, <u>regardless</u> of the velocity of its source or of its observer (thank you, Mr. Einstein). <u>Basic physics</u>.

So then, given $60 \times 60 = 3,600$ seconds per hour, light travels $3600 \times 186,282$, or roughly 671 million miles per hour. <u>Basic math</u>.

Our Sun is roughly 93 million miles from Earth...on the average, since Earth's orbit is slightly elliptical, not exactly circular. So then, on the average, it takes 93 mm divided by 671 mm or .1386 hours (so times 60) or 8.32 minutes for light from the Sun to reach Earth. And...

Radio waves travel at the same speed as light (waves) do...again, 186,282 miles per second.

The late Carl Sagan was the chief designer for and, in effect, oversaw the construction of *Voyagers I* and *II*, both space vehicles that were launched back in 1977, now forty-five plus years ago. *Voyager II* passed by Neptune en route to Pluto back in 1989, so roughly twelve and a third years after its launch. Voyager II was traveling at ±34,355 (average) miles per hour…so rather slowly by 'celestial measure.'

Therefore, back in 1989, Voyager II was <u>then</u>:

- 12.3 × 12 = 147.6 <u>months</u> from Earth
- times 30.42 (average) days = 4490 <u>days</u> from Earth
- times 24 hours per day = 107,760 <u>hours</u> from Earth
- And at ±34,355 miles per hour, Voyager II was then ±37 billion <u>miles</u> from Earth.

We know as above that light and radio waves travel at ±671 million miles per hour. So then at 37 billion miles away, a radio signal sent from Earth took 37 billion divided by 671 million, or roughly 5.51 hours, to reach *Voyager II*, then one way back in 1989. And this matters because at that time and distance…

A round-trip radio command out to *Voyager II* and then back to Earth took 5.51 + 5.51 = 11.02 hours (so 11 hours and 1.2 seconds) to confirm…that is to travel out and back, both ways, at 186,282 mps!

Today, *Voyager II* is out of touch, somewhere past Pluto, so it is known to be for sure the first manmade object to exit our solar system. Do recall *Star Trek's* episode, "V'ger," when another alien civilization allegedly found it! It is not known what became of *Voyager I* because its power failed long before *Voyager II's*. Science '<u>faction</u>' (a combination of 'fact' and 'fiction'), perhaps? It's okay to 'think about' these things!

Back to 1989. Dr. Sagan was given the honor of relaying the last possible earthly command to *Voyager II*. Reason: At 37+ billion miles away and with an 11+ hour radio turnaround time, the JPL's (Jet Propulsion Laboratory's) radios themselves were just barely able to communicate at all. Recall, *Voyager II* had been 'flying away' at roughly 35,000 miles per hour for roughly 12.3 years through space in temperatures well below -200° Fahrenheit (so negative). Both *Voyager*'s electronics and JPL's 'aim' had become marginal at best.

Dr. Sagan's command for *Voyager II* was for it to rotate 180° and then to operate its camera in reverse direction, pointing back from where/whence it had come. This command was received, obeyed, and confirmed. JPL received the images, put them to/through its computers, and amidst thousands of tiny points of light (primarily stars), was remarkably able to identify Earth (our planet). Due to its primarily water-laden surface (70-plus percent), Earth appeared as a very "Pale Blue Dot." And this was/is the title of Dr. Sagan's 1994 book—which again, is recommended reading.

Specifically, Dr. Sagan's chapter one, pages 6 and 7, paragraphs 3 through 6, as noted, are copied and attached. These paragraphs are Sagan's first response, his first impression upon viewing *Voyager II*'s actual photograph of Earth taken from ±37 billion miles out in space, but then with Voyager's camera still just within our solar system.

Figuring our Sun and our solar system is perhaps one of billions in our resident galaxy, the Milky Way, and then that our galaxy itself is one of billions in our known universe, this does make Earth a rather insignificant "Dot" in the collective cosmos...yes?

Last mathematics (I promise for now):

- Given ±70 billion stars (best estimate) just in the Milky Way (our galaxy), and...
- Given ±500 billion possible galaxies (again, best estimate) in the known universe, then...

- One must consider 70 × 500 or 35 sextillion possible (probable) galaxies in the cosmos.
- So then if Earth is one of eight planets (Note: There were nine, but Pluto has been de-listed—now more like a larger, icy asteroid), then 8 × 70 billion (possible suns/solar systems) equals 5.6 trillion possible planets that may likely exist just in our Milky Way.
- Note: Earthbound telescopes have now 'confirmed' several hundred of them...some "possibly life-friendly."
- So, if even one planet in 10,000 in our galaxy has liquid water (considered to be a 'life-friendly' requirement), then...
- 5.6 trillion divided by 10,000 equals
- 560 million possible planets, with possible life, just in our Milky Way.

So what? These numbers are beyond reasonable comprehension. Fair enough. But...

Are we "Earthlings" to assume that <u>we are the only ones intelligent enough to do this math</u>? <u>Anywhere</u>? Hard to say.

To Dr. Sagan's "Pale Blue Dot" analogy again, specifically that *Voyager* photographed...he saw a spot (again, of pale blue), which appeared to be "...a mote of dust suspended in a (so, only in <u>our</u>) sunbeam," and this does create a sort of interesting perspective, perhaps a parameter upon which we might consider:

1. the current and continuing man-made pollution of our atmosphere, which might lead to...
2. uncontrollable global warming and therefore the advisability of...

3. continued production, let alone…
4. the continued combustion of fossil fuels?

And this might be a reason to redirect certain oil companies in a bit of their future endeavors?

Accordingly, how about the future of any acceptable 'life' on our planet, regardless of the future possible 'climate,' when we simply run out of any recoverable coal, oil, or natural gas, so…

5. Does it matter?…or…
6. Should we care about what we are or are not doing now? Or later?

Even if the math is off by a factor of 10,000. Again, if we occupy either one of 2.8 sextillion or 2.8 quadrillion possible locations of possible life forms, what are all the following 'worth' to us, in the long term, regarding the future viability or even future existence of <u>our</u> life form here comfortably upon our actual "Pale Blue Dot"? So just currently, what should we care about:

A. ISIS, and/or the Taliban, and assorted 'other extremists'
B. Messrs V. Putin and Kim Jong Un…for some particular, current examples
C. Iran getting 'nuke(s)'
D. Coal-fired energy plants…here or in China
E. Internal-combustion engines…most anywhere
F. Deforestation of the Amazon Basin…for <u>any</u> purpose(s)
G. Mining tar sands in Canada and/or…

H. Transporting their contents to Texas or Louisiana to be...

I. Refined and sold...anywhere...and so to fuel any future...

J. Wars, in general, that create...

K. Millions of refugees...

L. Ukraine's ultimate border...

M. and on...to name a few concerning 'issues,' again, currently at hand.

One <u>could go</u> on and on and on. Here, upon our 'mote of dust,' what <u>are</u> we doing? What are we (or are we not) thinking? We now have 7+ billion of our life forms living upon our 'Dot.' We soon may multiply to 10+ billion, then to 15+ billion. How then will it be possible for us, *Homo sapiens*, to survive, <u>if</u> we do?

So once again, if you have not already done so, please read Dr. Sagan's four paragraphs, as follow...

Any thoughts? Any comments? Anytime?

Best regards, J.B.

As attached...

Carl Sagan from *Pale Blue Dot*, 1994:

But for us, it's different. Look again at that dot. That's here. That's home. That's us. On it everyone you love, everyone you know, everyone you ever heard of, every human being who ever was, lived out their lives. The aggregate of our joy and suffering, thousands of confident religions, ideologies, and economic doctrines, every hunter and forager, every hero and coward, every creator and destroyer of civilization, every king and peasant, every young couple in love, every mother and father, hopeful child, inventor and explorer, every teacher of morals, every corrupt politician, every "superstar," every "supreme leader," every saint and sinner in the history of our species lived there-on a mote of dust suspended in a sunbeam.

The Earth is a very small stage in a vast cosmic arena. Think of the rivers of blood spilled by all those generals and emperors so that, in glory and triumph, they could become the momentary master of a fraction of a dot. Think of the endless cruelties visited by the inhabitants of one corner of this pixel on the scarcely distinguishable inhabitants of some other corner, how frequent their misunderstandings, how eager they are to kill one another, how fervent their hatreds.

Our posturings, our imagined self-importance, the delusion that we have some privileged position in the Universe, are challenged by this point of pale light. Our planet is a lonely speck in the great enveloping cosmic dark. In our obscurity, in all this vastness, there is no hint that help will come from elsewhere to save us from ourselves.

The Earth is the only world known so far to harbor life. There is nowhere else, at least in the near future, to which our species could migrate. Visit, yes. Settle, not yet. Like it or not, for the moment the Earth is where we make our stand.

And the questions remain:

1. Are "we make(ing) this (our) stand?"
2. And if not: "How will we?" And: "When will we?"

Thank heaven for our grandchildren!

About The Author
(as of July 2022)

John Beaupré was born in Walla Walla, Washington. He grew up and was educated in Medina (Bellevue), Washington, well enough to have been accepted at both Stanford University and Dartmouth College. He chose Stanford and enrolled in 1959 as a 'pre-med' student, but became interested in other studies, to include the 'dismal' science of economics. After a break from college for a stint in the United States Army Special Forces, he graduated in 1984, well out of schedule. After another, almost totally unrelated career in the restaurant industry, John then became most interested in the decidedly 'less dismal' and therefore 'pure' sciences of physics and cosmology.

His interests, spurred on by exchanges with the late Dr. Carl Sagan of Cornell University, focused his attentions upon the very obvious but yet unanswered questions of: "Why us? Why are we, Homo sapiens, here? As we are? When we are? Where we are?" John seeks to find these answers in/within the actual mechanics of our human existence, as explained by the 'pure' (i.e. non-dismal) sciences…and logic.

Fizzicks and Cosmology 101 is his first effort. His next volume, Fizzicks 202, is in process and is intended to be published in a year or so (in relative disorder!).

John lives with Sharon, his wife of almost fifty-four years, and with his library and his pens and notes and papers in Santa Fe, New Mexico. He does most all of his research and his writing in and from print, and by his preference, not 'on-line.' John Beaupré may be contacted at: PO Box 1837, Santa Fe, NM 87504-1837.